LEARN TO **LEAD**

CIVIL AIR PATROL CADET PROGRAMS

CIVIL AIR PATROL USAF AUXILIARY

LEARN TO **LEAD**

CIVIL AIR PATROL CADET PROGRAMS

> "'He who is greatest among you shall be your servant.'
> That's the new definition of greatness."
>
> REV. DR. MARTIN LUTHER KING JR.

> "I don't mind being called tough...
> It's the tough guys who lead the survivors."
>
> GEN CURTIS LEMAY, USAF

> "Being powerful is like being a lady.
> If you have to tell people you are, you aren't."
>
> MARGARET THATCHER

> "A leader is a dealer in hope."
>
> NAPOLEON

> "Above all, do not
> lie to yourself."
>
> FYODOR DOSTOEVSKY

LEARN TO LEAD
Published by Civil Air Patrol
Maxwell Air Force Base, Ala.
3rd printing, Fall 2012

CURT LAFOND
with **NEIL PROBST** Associate Editor

LCMDR **KEVIN DAVIS**, USN
Former Cadet, Naval Aviator & Blue Angel
In Memoriam, 1975-2007

LEARN TO **LEAD**

CIVIL AIR PATROL CADET PROGRAMS

CONTENTS

CHAPTER 4
THE CADET NCO & THE TEAM

PROFESSIONALISM IS NOT THE JOB YOU DO, but how you do it. When there are new recruits to train, the armed services turn to the non-commissioned officer corps. Indeed, NCOs are experts in drill, the uniform, and fitness, but even more their professionalism makes a lasting impression. Air Force pilots know their aircraft are mission-ready because they trust in the professionalism of the NCOs who maintain them. When a fighter jet is low on fuel, it will be an NCO committed to an ideal of professionalism who refuels it from an altitude of 30,000 feet. Even when the Air Force needs to transform officer trainees into lieutenants, the NCOs' professionalism makes them ready for the challenge.

PROFESSIONALISM

OBJECTIVES:

1. Explain what "professionalism" is.
2. Defend the idea that leaders must be "professionals."

A leader is a professional. A leader strives to conduct himself or herself with a special quality called *professionalism.* What does this mean?

In the everyday sense of the word, "a professional" is simply someone who is paid for their work. In truth, professionalism requires much more.

First, *professionals must have a habit of putting the community's interest above their own.* The core value of "volunteer service" shows that CAP members think of themselves as professionals. Second, *a professional is someone who has special skills.* Their knowledge, experience, and competence in their field set them apart from others. Third, *professionals hold themselves and their peers to an ethical code.*[1] They practice their profession in a way that respects moral principles.

Leaders, especially military officers and non-commissioned officers, believe they meet all three criteria. Therefore, they strive to lead by example and display that special quality called professionalism.

CHAPTER OUTLINE
In this chapter you will learn about:

CHAPTER GOALS

1. Understand the role of the NCO and appreciate the importance of professionalism.

2. Describe leadership principles of concern to first-line supervisors.

3. Develop an understanding of team dynamics.

Professionalism.
It's a special quality leaders possess. It may be difficult to define, but is easy to spot.

STANDARDS

OBJECTIVE:

3. Explain what a "standard" is.

Standards. The best leaders have high standards. At a luxury hotel, you can expect to receive a high standard of service. A friend may tell you not to buy a certain kind of car if its quality is substandard. Michael Phelps set a new standard for athleticism by winning seven gold medals at the Olympics. We often speak of high standards and low standards. What is a standard?

> **"The leader's example is the most important standard of all."**

A standard is an established requirement, a principle by which something can be judged.[2] Put simply, a standard is like a yardstick or benchmark. Standards let people know what is expected of them. They help people understand what counts as acceptable or inferior work.

"Line Six, Sir!"
The sentinels who stand before the Tomb of the Unknown Soldier live by a creed. Its sixth line instructs them on the Old Guard's standard. It reads, "MY STANDARD WILL REMAIN PERFECTION."

It is vital that leaders set clear standards and communicate them to the team. In the military, standards are found in regulations, in special documents called technical orders, and in training manuals. Commanders can also establish standards orally, simply by declaring them to the team. Often, teams set their own informal, unofficial standards, the unwritten rules teammates must follow to be accepted by the group.[3] Even more importantly, how a leader acts sets the standard. *The leader's example is the most important standard of all.*

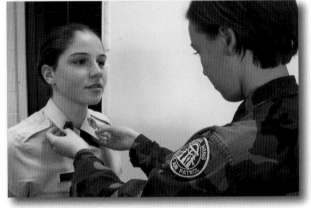

Not always "black and white," standards can vary depending on the situation. For example, your parents may allow you to dress casually for dinner, but if special guests are visiting, perhaps that standard is raised. Wise leaders learn how to make standards meaningful without allowing them to become so inflexible as to be impractical.

Meeting the Standard.
Two cadet NCOs double-check one another's uniforms to ensure each is meeting the standard.

Standards give leaders a way to express to the team what is expected from them.

PROFESSIONALISM IN ACTION Some examples of how you can show your professionalism

★ Checking your uniform and your airmens' uniforms frequently

★ Re-reading a chapter you studied long ago before teaching it to new cadets

★ Telling a fellow cadet that what she posted online is inappropriate

★ Sending a thank you note to someone who went out of their way to help you

★ Using downtime at a bivouac to check your gear before a hike

★ Surfing the web for helpful tips on public speaking or some other leadership topic you're weak in

THE NON-COMMISSIONED OFFICER

OBJECTIVES:

4. Discuss the challenge of transitioning from airman to NCO.
5. Describe seven major responsibilities of the NCO.

Air Force non-commissioned officers epitomize the Core Values. They have several duties, but if one is more important than the rest it is this: to lead by example. The challenge for the cadet NCO is to carry on this tradition.

> "New cadet NCOs have to transition from one who has been cared for to one who cares for others."

Making the switch from airman to sergeant can be difficult. *It involves transitioning from one who was cared for to one who cares for others; from one who was taught to one who teaches.*[4] Further, in the military, NCOs have authority to issue lawful orders to their people. As one expert said, "Rank does not confer privilege or give power. It imposes responsibility."[5]

Creed.
A formal system of belief intended to guide someone's actions.

CREED OF THE NON-COMMISSIONED OFFICER

There is no creed officially adopted by the U.S. Air Force for its NCOs, but the creed below is widely accepted and is based upon a creed used by the U.S. Army.[6]

No one is more professional than I. I am a Noncommissioned Officer, a leader of people. I am proud of the Noncommissioned Officer Corps and will at all times conduct myself so as to bring credit upon it. I will not use my grade or position to attain profit or safety. Competence is my watchword. I will strive to remain tactically and technically proficient. I will always be aware of my role as a Non-commissioned Officer. I will fulfill my responsibilities and display professionalism at all times. I will strive to know my people and use their skills to the maximum degree possible. I will always place their needs above my own and will communicate with my supervisor and my people and never leave them uninformed.

I will exert every effort and risk any ridicule to successfully accomplish my assigned duties. I will not look at a person and see any race, creed, color, religion, sex, age, or national origin, for I will only see the person; nor will I ever show prejudice or bias. I will lead by example and will resort to disciplinary action only when necessary. I will carry out the orders of my superiors to the best of my ability and will always obey the decisions of my superiors.

I will give all officers my maximum support to ensure mission accomplishment. I will earn their respect, obey their orders, and establish a high degree of integrity with them. I will exercise initiative in the absence of orders and will make decisive and accurate decisions. I will never compromise my integrity, nor my moral courage.

I will not forget that I am a Professional, a Leader, but above all a Noncommissioned Officer.

RESPONSIBILITY & THE NCO

What are the responsibilities of the non-commissioned officer?[7]

Epitomize the Core Values. NCOs must show by example that they are truly committed to integrity, service, excellence, and respect. They are charged with demonstrating superb military bearing, respect for authority, and the highest standards of dress and appearance.

Guide, Instruct, and Mentor. The NCO is a first-line supervisor, someone who ensures the junior members of a team accomplish the mission. To do that, they generously share their experience and knowledge.

Support the Leader. Although NCOs are leaders by virtue of their seniority, they still have bosses of their own. NCOs backup their leaders by enthusiastically supporting, explaining, and promoting their leaders' decisions.

Reward People. As the leader who is working closest with the troops, the NCO has a duty to recognize the hard work of people on their team. When they catch people doing things right, they praise them and point to them as role models.

> **Military Bearing.**
> How those in uniform carry themselves; bearing includes physical posture, mental attitude, how faithfully customs and courtesies are rendered, etc.

> **First-Line Supervisor.**
> A leader who oversees entry-level people; the lowest ranking member of a leadership staff.

Correct People. Because a leader's first duty is to accomplish the mission, if an NCO sees someone going about their job in the wrong way, he or she steps in

"NCOs step in and correct cadets in a helpful way."

and corrects the person in a helpful way. This also means NCOs speak up when they spot a safety hazard. The NCO corps takes their duty to protect airmen's safety very seriously.

Training Them Right. Cadet NCOs need to be great instructors and experts in all facets of cadet life.

Career Counsel. NCOs use their wealth of experience to help airmen through career counseling. NCOs tell airmen what opportunities are available to them in their organization, be it the Air Force or CAP. They steer their people toward activities that will help them advance and meet their personal goals.

Keep Learning. As a professional, the NCO is continuously trying to learn more about their specialty or career field. For cadet NCOs, this means learning more about leadership, aerospace, fitness, and character, in addition to the academic subjects they are studying at school.

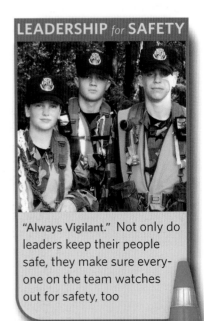

LEADERSHIP for SAFETY

"Always Vigilant." Not only do leaders keep their people safe, they make sure everyone on the team watches out for safety, too

NCO READINESS

OBJECTIVES:

5.   Identify the Air Force's three requirements for NCO readiness.
6.   Give examples of how Air Force NCO readiness standards apply to CAP cadets.

How do NCOs prepare to fulfill their many responsibilities? The Air Force identifies three areas of readiness.[8]

Technical Readiness. ***First, NCOs must be technically ready to accomplish the mission.*** This relates back to the "Leader as Expert" section in the previous chapter. The Air Force rightly insists that NCOs know the details of their job. What is the mission? How do we perform those tasks? What are the standards – how do we know if we've done a "good job"? What does the regulation require? These are some of the questions NCOs ask themselves to ensure they have the technical know-how to lead others.

Physical Readiness. ***Second, NCOs must be physically ready to accomplish the mission.*** If a leader is not healthy, he or she could hinder the team instead of helping it. Because of this requirement, NCOs are traditionally the ones who lead their units in fitness training. Again, leadership by example is expected. If NCOs do not exercise regularly and meet fitness standards, who else will?

Mental Readiness. ***Third, if an NCO's mind is not totally focused on their professional duties, the mission will suffer.*** This requires NCOs to effectively manage their stress, as discussed in chapter two. Mental readiness includes having healthy attitudes toward school and home life. It means being alert for signs of drug abuse and depression (for example, talk of suicide) in oneself and others. Cadet NCOs can show a commitment to mental readiness by promoting the wingman concept (see chapter two) and reaching out for adult help when life seems to be spinning out of control.

Mental Readiness. "Readiness" includes living the drug-free ethic. No wonder CAP cadets train with special goggles that simulate the effect of alcohol.

THE NCO'S LEADERSHIP TOOLKIT

SERVANT LEADERSHIP

OBJECTIVES:

7.   Define "servant leadership."

8.   Discuss why leaders should be servants first and leaders second.

9.   Give examples of servant leadership in action.

10.   Defend the idea that servant leadership is consistent with Air Force values.

Leadership is not about controlling people, but serving them.[10] This is one of the core beliefs of a leadership philosophy known as servant leadership. Simply defined, *servant leadership is when the leader sees himself or herself primarily as a servant of the team.* The goal of servant leadership is to enhance the growth of individuals in the organization and increase teamwork and personal involvement. The leadership theorist who first coined the term "servant leadership," introduced it as follows:

> *"The servant-leader is servant first...* It begins with the natural feeling that one wants to *serve*, to serve first. Then a [willful choice] brings one to aspire to lead. That person is sharply different from one who is *leader* first, perhaps because of the need to [fulfill] an unusual power drive or to acquire material possessions..."[11]

> "Leadership is not control, but service."

TO **SERVE** ...   *or* **TO BE SERVILE?**

A *servant* chooses to help people and is giving. Anybody can serve, so anybody can lead, even our youngest cadets.

But being *servile*, on the other hand, means to be treated like a slave. While others enjoy a life of leisure, your work never ends.

Leaders choose to serve because they realize there's nothing demeaning about being a servant.

Nicolaes Maes
The Idle Servant, 1655
The Netherlands

SERVANT LEADERSHIP AND THE NCO

Earlier in this chapter, we discussed the challenge of switching from airman to NCO, from one who is cared for to one who cares for others. Servant leadership is an important concept for new NCOs because it can help them make that transition.

It is easy for new leaders to become arrogant, to show off their rank and delight in bossing people around. (Do people who are full of themselves inspire you or turn you off?) So much of what we think we know about leadership is based on old concepts of power, not on the leader's potential to help individual people and the team succeed. **Servant leadership, then, is the new leader's vaccine against becoming self-centered or a bully.** It focuses the new leader on the needs of the team.

Tough, but a Servant.
As a servant leader, perhaps this first sergeant is thinking, "I want to help these cadets surpass my high standards." But she is not doing the hard work for her airmen, nor is she a bully. Servant leaders help individuals and the team grow.

EXAMPLES OF SERVANT LEADERSHIP

For example, an NCO informed by the idea of servant leadership will not use his or her rank to take a position first in line to eat, but rather will eat only after the team has been fed. The airmen come first. When conducting a uniform inspection, the servant-leader's goal will be to help each individual meet CAP's high standards, not try to intimidate the airmen or play "gotcha." **Servant leadership is not about a personal quest for power, prestige, or material rewards.**

The history of Christianity gives us a famous example of servant leadership that aspiring leaders from all backgrounds can appreciate. In the Bible, Jesus tells the apostles,

> "You know that among the Gentiles those whom they recognize as their rulers lord it over them, and their great ones are tyrants over them. But it is not so among you [the apostles]; but whoever wishes to become great among you must be your servant, and whoever wishes to be first among you must be slave of all."[12]

Humility as Virtue.
Not even a slave could be made to wash another's feet. Here, Christ does just that as he teaches about service.

SERVANT LEADERSHIP IN THE MILITARY

In an organization that has a top-down hierarchy, like the military with its formal chain of command, is it possible for rank to be emphasized too much? The "I order you to..." approach is easy to take when a leader literally has the power to imprison those who disobey. *Pulling rank is often seen as a lazy, immature, and counter-productive way to lead.* Proponents of servant leadership would advise military officers and NCOs to see themselves as servants first and authoritarian commanders last.[14]

A proper reading of the NCO Creed says as much, as does the Core Value "Service Before Self." Therefore, *servant leadership, and the idea of caring which it is built upon, is a natural fit for the Air Force.* "Caring bonds us together," according to a former Chief Master Sergeant of the Air Force. "When caring is lacking... mission failure is a very real possibility."[15] Air Force newspapers are filled with essays from commanders endorsing servant leadership.

Not only is servant leadership a wise approach to leading, everyone has the potential to be a servant-leader. As Martin Luther King once said, "Everybody can be great, because everybody can serve."

THE LEADER, THE WORKER, & THE BYSTANDER

Picture a country road leading through a wilderness to a river. A dozen soldiers are trying to build a bridge, but there are not enough men for such a challenging task. It's 1776 and the Revolutionary War is underway.

Now comes on a beautiful stallion an impressive, serious looking man. There is something powerful about the way he carries himself. He commands respect.

"You don't have enough men for the job, do you?" asks the man on horseback.

"No, we don't," answers the lieutenant in charge of the work detail. "The men will need a lot more help if we are to finish the bridge on time."

"I see," replies the man on horseback. "Well, why aren't *you* helping them? You are just standing back and watching them work."

"That, sir, is because I am an *officer*!" snaps the lieutenant. "I *lead*, I don't *do*."

"Indeed." The man on horseback then dismounts, tosses aside his cap, and rolls up his sleeves. He labors with the men under the hot sun for several hours until at last, drenched in sweat, he proclaims the job done.

"Lieutenant," says the man as he mounts his horse and prepares to depart, "the next time you have too much work and not enough men, the next time you are too important or high ranking or proud to work, send for the Commander in Chief and I will come again."

It was General Washington.[16]

COACHING & MENTORING

OBJECTIVES:

11.   Define "coaching."
12.   Explain how coaching relates to servant leadership.
13.   Discuss the elements of successful coaching.
14.   Give examples of some techniques to use during coaching's dialogue phase.

In chapter two, we discussed mentoring: how your mentor can help you and how as an aspiring leader you need to become "mentor-ready." The next step is to consider mentoring (or coaching, as we'll call it in this chapter – it is not yet time to distinguish between the two terms) from the perspective of the leader.

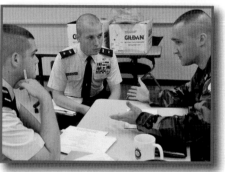

COACHING DEFINED

In leadership, ***coaching is the process through which leaders try to solve performance problems and develop their people.***[17]

Coaching is a person-to-person experience, a relationship between an experienced person (the leader or coach) and an inexperienced person (the

> **"If it's not positive and helpful, it's not coaching."**

follower or coachee).[18] Experts in the field of coaching borrow principles of servant leadership by teaching that coaching requires a bond of trust. If it's not positive and helpful, it's not coaching. Coaching is all about providing someone with guidance and support because a good coach is a servant.

THE NEED FOR COACHING

When does someone require coaching? Anytime a leader identifies a need to help someone reach a higher level of effectiveness.[19] Perhaps an airman is having trouble executing an about face. Or a basketball player cannot let go of a feeling that she was fouled and now her anger is getting the best of her. When someone struggles to reach a new level of excellence, that might mean it's time for coaching.

ELEMENTS OF SUCCESSFUL COACHING

What does successful coaching look like? There are four key elements:[21]

Dialogue. *Coaching is marked by dialogue, a two-way conversation between the coach and the trainee.* The coach talks with and listens to the coachee to try to understand what is blocking that person from succeeding. Once again, the principles of active listening discussed in chapter two come into play.

Empowerment. One assumption about coaching is that people learn more when they figure things out for themselves. *Empowerment occurs when the person who has all the answers resists the urge to jump in and "fix" someone's problem for them.* Instead, the coach chooses to help that person discover the solution on their own. As the old saying goes, give a person a fish and they eat for a day; teach them how to fish and they feed themselves for a lifetime.

Action. *The dialogue between coach and coachee must produce something.* Talk is meaningless unless followed by action. The coach's goal is to get the trainee to act, to try to do something differently and thereby solve the problem.

Improvement. *Ultimately, the goal of coaching is to help the coachee reach a higher level of effectiveness.* Coaching is successful if there is some sign of improvement or progress. If a cadet reduces the number of gigs from eight to one on their uniform inspection, the coaching worked. If the cyclist's top speed declines by 5 miles per hour, the coaching failed.

Facta Non Verba. It's Latin for "deeds, not words." Coaching is successful if there is some sign of improvement. Talk is cheap. At the end of the day, the trainee has to perform. General Patton (above) understood this when he said, "I'm a soldier, I fight where I'm told, and I win where I fight." The cadets holding the National Cadet Competition trophies understood, too.

TECHNIQUES OF SUCCESSFUL COACHING

How do leaders go about coaching someone? Continuing the discussion begun in chapter three, we see that coaching (and the whole of leadership) is partly an art because it takes imagination. Coaching requires creative thinking. Coaching also has a scientific aspect because it's based on a study of human behavior. There seems to be a process to coaching that when followed produces results that are somewhat predictable.[21]

Observation. ***Coaching begins with observation.*** Leaders need to watch how their people perform. The best way to gain an understanding of someone's strengths and weaknesses is through direct observation. For the NCO, this means paying close attention to each cadet's performance on the drill pad, in the classroom, during uniform inspections, and the like.

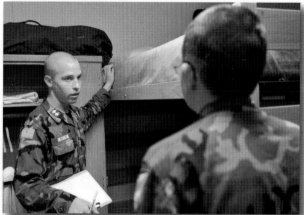

Attention to Detail.
New cadets learn the importance of paying attention to detail. But cadet NCOs and officers need to perfect that skill. By closely watching how their subordinates perform, leaders discover what areas need coaching.

Purpose. Second, before beginning a dialogue with the coachee, the coach needs to be clear about the purpose of that discussion. ***Rather than shooting from the hip, the coach should enter the dialogue having a plan.*** Of the many observations you've made about a particular cadet's performance, which one or two do you want to focus on? What is the coachee doing well? Which issues are most important? What will be the consequences if the coachee does not fix the performance problems? The coach must be ready to lead the dialogue.

Dialogue. Third is the dialogue itself. ***The dialogue is at the very center of coaching.*** There are several tactics a coach can use to lead a good dialogue.

    Mirroring. The coach uses words to paint a picture of the coachee's behavior. Figuratively speaking, he holds a mirror up to the trainee and asks what they see. The idea here is to get the coachee to identify their own strengths and weaknesses, to see themselves as others see them.

> **"Hold a 'mirror' up to the trainee and ask what they see."**

    Questioning. Asking open-ended questions – the kind of questions that cannot be answered with a simple "yes" or "no" – is a sure way to have a meaningful conversation. Open-ended questions empower the trainee to find answers to their own problems.

    Active Listening. The coach listens intently to what the trainee says and does not say. Attention is paid to the trainee's body language and the emotions that come to the surface. Through active listening, the coach develops a deeper understanding of the trainee's attitude, frustrations, and concerns.

Open-Ended Question.
A question that requires more than a "yes" or "no" answer.

EXAMPLES:

What did you enjoy most about your orientation flight?
NOT: Was your flight fun?

What preparations have you made for the bivouac?
NOT: Are you ready?

Validating. *Simply having someone acknowledge they understand what you are going through can help.* This approach to coaching is called validation. When you were little and skinned your knee, your mom said, "Yes, I know it hurts." She validated your feelings. Validation also allows the coach to point out something the coachee does not see in themselves.

Story Shifting. There are many ways to look at a given situation. *Through a "story shift," the coach asks the trainee to look at the problem in a different way or from someone else's perspective.* Story shifts help people realize there is more to an issue than they may first realize.

Addressing Fears. *Fear can stop a person from moving forward.* Fear is the reason people resist trying new things. A coach can help the trainee identify their fears and face them.

Freedom From Fear.
Good coaches help people overcome their fears.

Finding The Bottom Line. Some problems are so overwhelming we don't know where to begin. Even describing the problem can take more energy than we have. *Through the "bottom line" approach, a coach helps the coachee express the problem in one or two simple sentences.* The idea here is that a problem cannot be overcome until the trainee is able to define it precisely using everyday language.

Providing Direct Feedback. The coach gives feedback about something the trainee is doing. *To be effective, feedback needs to be positive, constructive (serving a meaningful purpose), and specific (with some concrete examples).* After the coach provides feedback, the trainee is asked to change how they're doing their job.

Follow-Up. The fourth and final step in the coaching process is the follow-up. Giving and receiving feedback is a critical part of coaching. Effective coaching includes follow-up that monitors how well the trainee is doing. Follow-up allows the coach to praise the trainee for working hard. It is also an opportunity to identify the next challenge that coach and trainee want to work on together.

8 COACHING TECHNIQUES

Mirroring

Questioning

Active Listening

Validating

Story Shifting

Addressing Fears

Finding the Bottom Line

Providing Direct Feedback

SUPERVISION & CONSTRUCTIVE DISCIPLINE

OBJECTIVES:

15. Defend the idea that trust and fairness are cornerstones of supervision.
16. Define the term "punishment."
17. Explain what "constructive discipline" is.
18. Discuss principles of constructive discipline.

One duty of a leader is to supervise the team, to ensure its members meet the standards. *To supervise means to observe and direct people in fulfillment of the mission.* What principles guide leaders in their capacity as supervisors?[22]

THE TRUSTING & FAIR SUPERVISOR

Trust. *Trust is a cornerstone of supervision.* It would be impossible for a leader to unblinkingly supervise every team member all day long. "You must trust and believe in people," wrote playwright Anton Chekhov, "or life becomes impossible." After all, a supervisor is a servant-leader, not a police officer waiting to catch a thug doing something illegal.

> "Trust and believe in people... Catch them doing things right."

Fairness. Likewise, fairness is a second cornerstone of supervision. *Fairness means following an impartial set of rules and applying them equally to everyone.* Fairness means not playing favorites. When team members believe they are being treated unfairly, they will be less willing to cooperate with their leaders. It is particularly challenging for a cadet NCO to lead fairly because among their subordinates might be their brother, sister, or best friend. Cadet NCOs have to work extra hard to avoid even the appearance of treating fellow cadets unfairly.

The Need for Wisdom. Everyone agrees that "being fair" is a virtue. *But distinguishing between fair and unfair requires wisdom.* A leader's decisions about fairness will be open to debate and second-guessing. "When I was a young coach," football's Bear Bryant reflected, "I used to say, 'Treat everybody alike.' That's bull. Treat everybody fairly." Bryant would have us believe everyone on his team is alike in that each is a football player, but each player has his own unique abilities and needs. Each comes from different circumstances. Each responds to the coach's leadership in his own way. *The concept of fairness asks leaders to treat things that are alike in the same way.* However, knowing which things are alike and which are unalike is a judgment call requiring wisdom.

The Bear's Coaching Wisdom. One of the all-time great coaches was college football's Bear Bryant. "When I was a young coach, I used to say 'treat everbody *alike*,'" said the Bear. "That's bull. Treat everybody *fairly*."

THE CONSTRUCTIVE DISCIPLINARIAN

In our culture, many confuse discipline with punishment. Recall that in chapter one we showed how the word discipline can be traced to the word *disciple*, which is a person who follows the instructions of their teacher. In contrast, **punishment is a negative consequence**. Given the choice between punishment and being left alone, you'll choose to be left alone and skip the punishment. **Therefore, punishment teaches someone only what behaviors to avoid. It does not teach someone what they actually should be doing.**[23]

Constructive discipline is a learning process that provides an opportunity for positive growth.[24] Leaders apply constructive discipline when followers are able but unwilling to meet the leader's standards. The goal is not necessarily to punish someone but get them back on course so they meet the standards. Discipline then can be positive. How do leaders apply constructive discipline?[25]

Praise in Public
Most people like being recognized for their hard work. A handshake and a "congrats" given in front of the whole team goes a long way, especially in a volunteer organization like CAP.

Know that Ability Differs from Willingness. An airman walks past the squadron commander, while outdoors, without saluting. Is that failure a result of the airman not knowing the standard? Or does the failure suggest the airman is unwilling to show respect for the officer? Before beginning to apply constructive discipline, the leader first needs to verify the facts. **There are many reasons for people to fall short of a standard.** Perhaps they are confused and need extra training. Maybe their trainer did not do a good job explaining the subject. Once again, the leader would do well to remember that trust and fairness are at the heart of being a good supervisor.

Praise in Public, Correct in Private. **This is one of the fundamental laws of leadership.** Most of us are pleased to receive praise in front of our peers, but who enjoys being reprimanded before an audience? A public dressing down is more apt to breed resentment than improvement. Sincere appreciation for a job well done is an easy, cheap, and amazingly effective form of motivation. Public praise also reinforces good behavior. It fosters a healthy sense of competition. When you are praised for doing a good job, your teammates will want to do likewise.

Collect Them All.
Challenge coins are a popular way to say "great job."

Choose the Right Time. Some leaders are like the ostrich. They stick their heads in the sand and hope problems go away. The longer a leader waits to use constructive discipline, the worse the problem becomes. ***Constructive discipline must take place when the problem behavior is still fresh in the follower's and the leader's mind.*** Would you feel helped by a leader who delayed telling you that you were doing something wrong? Would you respect a leader who stored up your short-comings and fired them at you weeks or months after they happened? If leaders step in to correct people right away, they can calmly deal with one problem at time before the situation gets out of control.

Control Emotions. ***When disciplining a subordinate, a leader stays calm.*** "When angry, count to ten before you speak," advises Thomas Jefferson. "If very angry, a hundred." But when a capable follower chooses not to respect the team's rules and standards, it is appropriate for a leader to change their demeanor, their outward attitude. The experts say it is time to "raise the emotional content to a moderate level." What does that mean? The smile disappears from the leader's face. Their tone of voice signals that their message is important. The leader stands up straight and looks the person in the eye. The leader might let what they've said sink in with a few moments of silence. It's a time to be serious. On the other hand, constructive discipline does not involve shouting or getting angry. ***The leader never loses control.*** No attempt is made to frighten, intimidate, or humiliate the other person. Emotions show people what we are feeling. Going from an everyday attitude to a "no nonsense" stance signals a change in the leader's attitude. That change also teaches followers that they need to make some changes as well.

Focus on Performance. Leaders are not bullies. ***To keep constructive discipline positive, leaders focus on performance.*** They don't attack their people personally. For example, a leader may say, "Your repeated failure to get a haircut is unacceptable. We're cadets and we are to wear the uniform properly." But a leader would not say, "I hate you because you're a long-haired scrub." For criticism to be constructive, it must be specific, precise. It is no time for generalities. Effective supervisors focus on performance. ***They criticize bad behavior and inanimate objects, not the offending individual.***

HOW TO MAKE A "DISCIPLINE SANDWICH"

Positive feedback (bread)

Constructive criticism (cheese)

Positive feedback (bread)

EXAMPLE:

The Bread. Cadet Curry, you've been doing a great job with your uniform. I can tell you're proud to wear it. Your appearance brings credit to our squadron.

The Cheese. But I'm concerned about your frequent swearing. That doesn't bring credit to our team. We've talked about this before, and you told me you know that's not acceptable. If you're to remain an element leader, we absolutely need you to lead by example. No more swearing. Is that understood?...

The Bread. Good. Now keep working hard. You've got tons of potential and I want you to succeed.

"Criticize the bad behavior, not the individual."

21

GENERAL LEE REPRIMANDS A FRIEND

from **THE KILLER ANGELS** by Michael Shaara

He saw a man coming toward him, easy gait, rolling and serene, instantly recognizable: Jeb Stuart. Lee stood up. This must be done. Stuart came up, saluted pleasantly, took off his plumed hat and bowed.

'You wish to see me, sir?'

'I asked to see you alone,' Lee said quietly. 'I wished to speak with you alone, away from other officers. That has not been possible until now. I am sorry to keep you up so late.'

'Sir, I was not asleep,' Stuart drawled, smiled, gave the sunny impression that sleep held no importance, none at all.

Lee thought: here's one with faith in himself. Must protect that. And yet, there's a lesson to be learned. He said, 'Are you aware, General, that there are officers on my staff who have requested your court-martial?'

Stuart froze. His mouth hung open. He shook his head once quickly, then cocked it to one side.

Lee said, 'I have not concurred. But it is the opinion of some excellent officers that you have let us all down.'

'General Lee,' Stuart was struggling. Lee thought: now there will be anger. 'Sir,' Stuart said tightly, 'if you will tell me who these gentlemen...'

Gen Robert E. Lee, CSA

Maj Gen J.E.B. Stuart, CSA

> **"General Stuart," Lee said slowly, "you were the eyes of this army."**

'There will be none of that.' Lee's voice was cold and sharp. He spoke as you speak to a child, a small child, from a great height. 'There is no time for that.'

'I only ask that I be allowed – '

Lee cut him off. 'There is no time,' Lee said. He was not a man to speak this way to a brother officer, a fellow Virginian; he shocked Stuart to silence with the iciness of his voice. Stuart stood like a beggar, his hat in his hands.

'General Stuart,' Lee said slowly, 'you were the eyes of this army.' He paused.

Stuart said softly, a pathetic voice, 'General Lee, if you please...' But Lee went on.

'You were my eyes. Your mission was to screen this army from the enemy cavalry and to report any movement by the enemy's main body. That mission was not fulfilled.'

Stuart stood motionless.

Lee said, 'You left this army without word of your movements, or of the movements of the enemy, for several days. We were forced into battle without adequate knowledge of the enemy's position, or strength, without knowledge of the ground. It is only by God's grace that we have escaped disaster.'

'General Lee.' Stuart was in pain, and the old man felt it, but this was necessary; it had to be done as a bad tooth has

to be pulled, and there was no turning away. Yet even now he felt the pity rise, and he wanted to say, it's all right, boy, it's all right; this is only a lesson, just one painful quick moment of learning, over in a moment, hold on, it'll be all right. His voice began to soften. He could not help it.

'It is possible that you misunderstood my orders. It is possible that I did not make myself clear. Yet this must be clear; you with your cavalry are the eyes of the army. Without your cavalry, we are blind, and that has happened once, but must never happen again.'

There was a full moment of silence. It was done. Lee wanted to reassure him, but he waited, giving it time to sink in, to take effect, like medicine. Stuart stood breathing audibly. After a moment he reached down and unbuckled his sword, theatrically, and handed it over

> **It had to be done as a bad tooth has to be pulled, and there was no turning away.**

with high drama on his face. Lee grimaced, annoyed, put his hands behind his back, half turned his face. Stuart was saying that since he no longer held the General's trust, but Lee interrupted with acid vigor.

'I have told you that there is no time for that. There is a fight tomorrow, and we need you. We need every man, God knows. You must take what I have told you and learn from it as a man does. There has been a mistake. It will not happen again. I know your quality. You are a good soldier. You are as good a cavalry officer as I have known, and your service to this army has been invaluable. I have learned to rely on your information; all your reports are always accurate. But no report is useful if it does not reach us. And that is what I wanted you to know. Now.' He lifted a hand. 'Let us talk no more of this.'

MOTIVATION

OBJECTIVES:

19. Define "motivation."
20. Explain why leaders must understand what motivates their people.
21. Distinguish between intrinsic and extrinsic rewards.
22. Defend the idea that the key to motivation is to communicate a shared purpose.
23. Defend the idea that leading volunteers is more challenging than leading paid employees.

Why did you join CAP? Why have you remained a cadet? Why are you reading this book when you could be spending your time doing something else? *Motivation is the reason for an action. Motivation is that which gives purpose and direction to a behavior. In short, motivation is your "why."*[26]

A person's why, their motivation, is their strong reason for desiring something. It is not the thing they desire, but the inspiration for it. For example, a cadet may desire a pilot's license. Why? The sheer thrill of flying may be their motivation. Another cadet may desire to become a lawyer. Why? Perhaps the cadet knows most lawyers are paid well and money is their motivation.

> **"What motivates people? Money, power, fear, honor..."**

Talent is different from motivation.[27] Someone may have enormous potential for success, but if they aren't motivated, they won't perform well. In contrast, someone who is motivated to do something is apt to stick with that task, even when it becomes difficult.

Leaders are concerned with motivation because it is what answers the question, Why should I pursue this goal? *Leaders who understand what motivates their people are apt to get them to fulfill the team's goals.*[28]

SOURCES OF MOTIVATION

What motivates people? The list never stops. Money, power, peer pressure, revenge, honors, fame, fear, competition, a sense of belonging, a desire to make a difference, prestige, hunger, a tough challenge, pride, personal achievement, status... this is a list that never ends.

Camaraderie?

Duty?

Excitement?

Competition?

RICHARD SIMMONS, PLEASE DON'T DIE!

If a motivational leader is someone who can convince you to achieve the impossible, then Richard Simmons is world-class.

Simmons, "the clown prince of fitness," is so flamboyant that some people are initially turned off by his special brand of charisma.

But he has an incredible talent for using laughter, outrageous costumes, and a wacky enthusiasm to connect with overweight people, many of whom are crippled by low self-esteem. Simmons inspires people to change. The man overflows with hope.

"I was handicapped all my life until I lost weight," he said. "I was always the first in line for lunch, and the last to be chosen for sports. I know how it feels."

As a teen growing up in New Orleans, his weight ballooned to 268 pounds. A weird fat kid, Simmons was an easy target for bullies.

But after an anonymous person left him a note reading, "Fat people die young. Please don't die," Simmons found his inspiration. He changed his life and decided to help others follow his lead.

Of special interest to Simmons are people who are "morbidly obese" – the heaviest of the heavy. Every day he personally calls or emails dozens of people who terribly despair about their poor health. Most of these people have rejected pleas from their family, friends, and doctors, but somehow Simmons' deep sincerity and unique personality enables him to change lives.

While experts debate solutions to America's obesity epidemic, Simmons' focus is simple: motivation. "Where we have missed the boat is the tapestry of motivation, eating and exercise," he says. "People ask me, how many days do you exercise? I ask them, how many days do you eat?"

Having helped Americans lose over three million pounds of fat, Richard Simmons is arguably one of the all-time great motivational leaders.[29]

Psychologists loosely group motivators into two groups:[30]

Intrinsic rewards *are motivators at work within you*. They drive you to do something because of how they make you feel. Maybe you go cycling just for the fun of it. Maybe you rake the leaves in your grandmother's yard not because someone assigned you that chore but because you simply want to help. The incredible amount of energy volunteers devote to community service shows how powerful intrinsic motivation can be.

Extrinsic rewards *are motivators at work outside of you*. They drive you to act because you receive something tangible in return. A paycheck is an external motivator. An employee works hard and at the end of the week is rewarded with money. From the leader's perspective, one shortcoming of extrinsic rewards is they can focus followers on the rewards (the money, the job title, the trophy) and not on the mission. Stop providing the reward and the follower may stop work on the mission.

Intrinsic.
Naturally belonging to something; essential.

Extrinsic.
Not belonging to something; a thing that comes from the outside.

SHARED PURPOSE

The key to motivation is to communicate a strong sense of shared purpose.[31] Leaders can motivate people by linking the team's mission with each individual's long-term goals. The most effective leaders try to motivate by gaining agreement, by appealing to shared values, by appealing to the follower's sense of what is right and what needs to be done.

LEADERSHIP IN A VOLUNTEER ORGANIZATION

Nowhere is the need to motivate by appealing to shared values more apparent than in a volunteer organization like CAP. Volunteers help perform the mission simply because they want to. They find something worthwhile, some intrinsic reward in their volunteer work.

If volunteers are unhappy, if they do not see how their personal goals align with the volunteer organization's goals, they can vote with their feet and leave the organization. In theory, paid employees can always leave their jobs, but that is often an impractical choice because the employees need their paychecks. Therefore, there's a greater need to meet people's needs and to lead with a shared purpose in mind in a volunteer organization.

Compared with leaders of paid employees, volunteer leaders will find fewer extrinsic rewards in their toolkit.[32] After all, leaders of volunteers cannot pay people more. When money is not available as a motivator, a leader has to work harder in finding other ways to influence people.

CAP cadets have an edge in their leadership training because they are learning to lead in a tough setting: a volunteer organization.

TOOLKIT FOR MOTIVATING VOLUNTEERS

Some tools for intrinsic motivation:

Allow the cadet to make a difference

Build esprit de corps

Help the cadet achieve a personal goal, such as an Academy appointment

Some tools for extrinsic motivation:

Publicly praising the cadet

Awarding the cadet a certificate, trophy, or ribbon

Promoting the cadet

Assigning the cadet to a prestigious staff position

ONE HERO AMONG THOUSANDS

How long have you been in CAP? Col Ben Stone was in longer. At the time of his death in 2009, Col Stone boasted the record for the longest continuous service to CAP – 68 years.

As a founding member of CAP, Col Stone trained our subchasers during WWII. He went on to serve in nearly every imaginable position, but the cadets were his biggest love.

"The young men and women in the cadet corps of CAP are our future leaders and need help in understanding their role in leading our country," he said.

But did he make a difference? America's first astronaut thought so, for Col Stone gave Alan Shepherd his first flight in an airplane.

A self-described "100 percent patriot who loves my God, my country, and my family," Col Stone, like so many other CAP members, lived the Core Value of Volunteer Service.

He was just one of the thousands of volunteer heroes motivated to serve despite *never receiving a paycheck.*[33]

Col Benjamin Stone, CAP

THE NCO / OFFICER RELATIONSHIP

OBJECTIVES:

24. Compare and contrast the duties common to NCOs and officers.
25. Identify the three levels at which leadership is practiced.
26. Identify the three levels of leadership skill.
27. Describe ways leaders can effectively support their boss.

The lesson of the chain of command is that everyone has a boss. As discussed in chapter one, even the president answers to the American people. Likewise, although NCOs find themselves in leadership roles, they remain accountable to their superiors. What can be said of the NCO / officer relationship?[34]

NCOs

1. Focus on short-term needs of individual cadets and a small team

2. Ensure people comply with policies, rules, and standards

3. Train people to do their jobs

4. Fulfill the goals of the team, carry out activity plans, and develop a sense of teamwork

5. NCOs get the job done

OFFICERs

1. Focus on long-term needs of the whole team or teams of teams

2. Establish policies, rules, and standards

3. Assign people to the right jobs

4. Set goals for the team, plan activities, and organize a team of leaders

5. Create the conditions necessary for the team to succeed

Cooperating With Cadet Officers. As a new cadet NCO, how is your relationship with your cadet officers?

Do you support them?

Do you follow their instructions?

Do you tell them what the airmen need?

Do you help ease their leadership burden?

THE LEADERSHIP MATRIX

The skills a leader needs depends on the level they are leading at. For example, a sergeant who often works one-on-one with airmen needs outstanding people skills. A general who leads a massive organization still needs people skills, but sophisticated planning and organizational skills are even more important. The Air Force uses a matrix (see right) to illustrate that as the level of leadership changes, so do the skills required.[35]

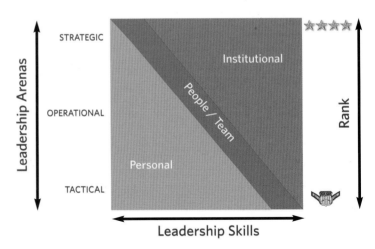

Slightly modified for CAP

LEADERSHIP ARENAS[36]

Strategic. ***The strategic arena is the highest level of leadership.*** Strategic leaders have responsibility for large organizations. They focus on the long term health of the institution. As such, strategic leaders are rarely involved in day-to-day operations. Using search and rescue as our example, strategic leaders ensure the organization has purchased the right type of aircraft. They see to it that the institution has the funding it needs to succeed. They try to imagine what challenges the organization will be facing in five or ten years.

Operational. ***The operational arena is the middle level of leadership.*** The challenges a leader faces here involve organizing and directing tactical-level leaders. They coordinate the minute details of the tactical teams performing the mission. They are middle managers who bridge the gap between the tactical and strategic. For example, in search and rescue, the mission base staff organizes people into aircrews and assigns them to search certain areas.

Tactical. ***The tactical arena is the lowest level of leadership.*** The challenges a leader faces here are immediate, small in scale, and relate to everyday tasks. That is not to say leadership at the tactical level is unimportant. On the contrary, it is where "the rubber meets the road." Tactical leaders are first-line supervisors who help the people who directly accomplish the mission. For example, the pilot leading an aircrew on a search mission is leading at the tactical level.

LEADERSHIP SKILLS[37]

Personal. Personal-level leadership skills involve leading oneself and leading others, especially in a one-on-one or small team setting. Volumes 1 and 2 of this textbook focus on personal leadership skills.

Team. Team-level leadership skills involve leading large teams by directing other leaders. Volume 3 of this textbook focuses on the skills of indirect or team leadership.

Institutional. The highest level of leadership skill involves leading an entire establishment (e.g. not just a fighter squadron but the entire U.S. Air Force). Volume 4 of this textbook introduces cadets to some perspectives needed to lead at the strategic or institutional level.

RESPONSIBILITY BIG & SMALL
A look at the three arenas of leadership

STRATEGIC: Big picture leadership for an entire institution

OPERATIONAL: Leading and coordinating teams of teams

TACTICAL: On the spot leadership where the rubber meets the road

SUPPORTING THE BOSS

As discussed earlier, NCOs are leaders and yet they still answer to higher-level leaders. What are some principles that guide them in remaining good followers?

Command Intent. *Command intent is the leader's concise expression of purpose.*[38] It describes what experts call the desired end state. In short, it explains the overall result a commander wants the team to achieve. No matter how thoroughly a leader explains the mission, words alone may not be enough. Effective leader/followers try to understand the command intent. When unforeseen problems arise on the job, an understanding of command intent guides the follower in solving the issue in a way the boss would find acceptable without the follower having to stop work, find their superior, and seek guidance.

Initiative. *Initiative is the ability to make sound judgments and act independently.*[39] Leaders who show initiative do a job because they see it needs to be done. They do not wait for tasks to be assigned to them. However, there is a fine line between taking initiative and doing one's own thing. For initiative to be helpful, it needs to be in agreement with the command intent.

Respectful Dissent. *Dissent in a military-style organization may seem contrary to the principles of discipline and following orders. But in fact, the Core Value of Excellence requires it.*[40] How do leaders dissent in a respectful manner? First, they use the chain. The chain of command is the solution, even when it is the problem. Second, they argue calmly and objectively. They explain why their idea works best for the team. Third, they are tactful. Frank discussions with the boss belong behind closed doors. Finally, no matter what the boss's final decision may be, a loyal leader/follower will be prepared to support that decision, so long as it is lawful and moral.[41]

Desired End State.
What a leader hopes to achieve; what the world will look like when the goal has been met

What Is The Command Intent? In big operations like encampments, the commander cannot explain how you are to carry out every single duty you are assigned. Therefore, good NCOs try to understand the "command intent."

Dissent.
To express an opinion that differs from the official view

> "Frank discussions with the boss belong behind closed doors."

CAN YOU HEAR ME NOW?

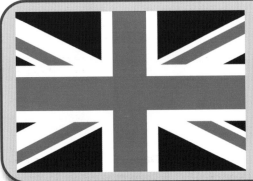

"...We shall not flag or fail. We shall go on to the end, we shall fight in France, we shall fight on the seas and oceans, we shall fight with growing confidence and growing strength in the air, we shall defend our Island, whatever the cost may be, we shall fight on the beaches, we shall fight on the landing grounds, we shall fight in the fields and in the streets, we shall fight in the hills; we shall never surrender."[42]

WINSTON CHURCHILL *expresses with absolute clarity his command intent upon becoming Prime Minister of the United Kingdom during WWII*

Completed Staff Work. *One truism about leadership is that one should never complain about a problem without offering a solution.* Subordinate leader/followers are obligated to provide the boss with "completed staff work."[43] In short, this means the subordinate must thoroughly examine all aspects of a problem before bringing it to the boss's attention. *They must coordinate their proposal with the other stakeholders – the people who have a direct or indirect interest in the issue.* Ideally, the proposed solution should be crafted in such a way that the boss need only say, "I agree, do it."[44] To do otherwise is to burden the boss's limited time. The principles of completed staff work are the antidote to the complaint, "my ideas are never considered."

Updates and Advice. *One of a leader's duties as a follower is to keep their superior informed of any issues he or she would want to know about.* This requires judgment.[45] Does my boss need to know about this news? Will my boss's superiors ask him or her about this issue? Would my boss want to discuss this issue before one of his subordinates acts? No leader wants to be surprised by bad news or find about it secondhand. Likewise, one role of a leader is to provide advice and direction. Leader/followers should remember that their superior is a resource for advice on how to approach the challenges they are facing.

When learning how to be a leader who works for yet another leader, the Core Values continue to be guideposts. *Do what is right and the boss will support you.* Forget integrity, service, excellence, or respect, and the boss will lose confidence in your ability to lead yourself, let alone others.

An Update for America.
According to Article II of the U.S. Constitution, the president must report to the Congress on the state of the Union. This annual address is perhaps the ultimate form of having to update the boss on key issues affecting the mission.

whine n. a long, high-pitched cry against the world's many horrible injustices, especially as they affect poor little you.

There's respectful dissent, and then there's whining. Before criticizing a problem, come up with a solution.

KEEPING THE BOSS INFORMED: THE CHECK RIDE SYSTEM

When a responsible leader accepts a challenge, he or she will keep the boss informed as to how the project is going. No one likes surprises. The boss will want to see that the project is proceeding as he or she envisioned.

If you show your boss that you're making progress in getting the job done and heed the boss's advice and redirection, you're sure to build trust. This timeline shows how cadet staff can enjoy some freedom as they plan projects on their own while still keeping their boss informed.

Scenario: A senior member asks a cadet NCO to teach a class at an upcoming squadron meeting.

1. Assignment. Senior or experienced cadet officer meets with the cadet to discuss goals and vision for the class or activity.

2. Preparation. Cadet begins to do some thinking and finds and personalizes a lesson plan. Cadet should rely on published lesson plans vs. original work.

3. Check Ride. Cadet presents their ideas in depth to the cadet officer or senior, who provides mentoring and quality control.

4. Execution. Cadet leads activity or class.

5. Feedback. Cadet seeks their supervisor's help. They work together to identify ways the cadet can improve next time.

| 2 Weeks Prior | 1 Week Prior | D-Day | 1 Week Afterward |

TEAM DYNAMICS

OBJECTIVES:

28. Describe the seven needs of a team.
29. Discuss common pitfalls that teams are susceptible to.
30. Describe the four stages in Tuckman's model of team dynamics.
31. Describe the "L.E.A.D." model.

If everyone believes they are an above average driver, why are there so many idiots on the road? According to one leadership expert, ninety-seven percent of managers believe they are skilled at leading teams. However, that expert also found that most people have experienced poor leadership at one time or another.[46] It seems that leading a team is a lot harder than it looks.

THE TEAM ENVIRONMENT

In chapter two, we discussed the benefits of teamwork. **T**ogether **E**veryone **A**chieves **M**ore. We also discussed what it takes to be an effective team member. Now we turn our focus toward what it takes to lead the team. A good place to begin that investigation is to ask, What do teams need to succeed?

SEVEN NEEDS OF TEAMS

Every team has certain needs simply because of the nature of teams. This is true regardless of the team's job or its setting. A football team, a flight of cadets, or a group of technicians building an airplane will each hold certain needs in common. Among those needs are the following:[47]

Common Goals. *A team must be organized around a common goal.* Without a goal or a mission, there is no need for the team to exist.

Leadership. *Every team needs leadership.* In chapter three, we learned that the Air Force defines leadership as "the art and science of influencing and directing people to accomplish the assigned mission." Therefore, teams need people who can move the team

7 NEEDS OF TEAMS
Common Goals
Leadership
Involvement from All
Good Morale
Open Communication
Mutual Respect
Ways to Resolve Conflict

Teamwork at 30,000 feet. How can aircraft refuel without teamwork? If both pilots and the boom operator work at cross-purposes, this B-52 will run out of fuel and fall from the sky.

toward its goals. Leadership can come from formal (official) leaders like squadron commanders, or informal (unofficial) leaders such as ordinary team members who have good ideas.[48]

Involvement of All Members. *The main idea of teamwork is to harness each individual's strengths.* Therefore, if even one member holds back their energy and talents, the team suffers.

Good Morale. Membership on a team is supposed to be a positive experience for all involved. Therefore, *team leaders are concerned with morale: the level of confidence, enthusiasm, and discipline of a person or group at a particular time.*[49] When the team succeeds and morale is high, the team develops a special quality called esprit de corps. Simply put, *esprit de corps is a sense of team pride, fellowship, and loyalty.*[50] Esprit is possible only when the team is effective in accomplishing its mission.

Open Communication. Team members need the ability to communicate with one another, with their leader, and with other teams. *Ineffective communication among team members and ineffective use of meeting time are the two biggest complaints people have about team leadership.*[51] Communication skills are discussed in depth in chapter eight.

Mutual Respect.   Once again the Core Value of Respect shows its importance. *Team members must show a commitment to mutual respect and all that it entails.* When someone plays favorites, fails to honor their commitments, or fails to respect the dignity of each individual on the team, the team is less able to fulfill its goals.

Fair Way to Resolve Conflicts. "Hell," one philosopher pronounced half-jokingly, "is other people."[52] *In a team environment, conflicts are always bound to arise. The challenge is to resolve them fairly and professionally.* It would be naive to think all conflict can be eliminated. Because conflict is never pleasant, it's no surprise that many leaders try to avoid it, sometimes at all costs.[53] Effective teams find ways to resolve conflicts, either informally (such as by people simply being trusting and brave enough to express their frustrations) or formally (such as through an official complaint process).

> **"Conflicts are bound to arise. The challenge is to resolve them fairly."**

CADET MORALE

Leaders monitor the team's morale. In a volunteer organization like CAP, morale is especially important because if morale plummets, people may choose to quit.

Surveys tell us cadet morale is highest when cadets have:

★ Well-Trained Adult Leaders

★ Opportunities to Fly

★ Lots of Exciting Weekend Activities

★ Freedom to Lead

★ Hands-On Activities During Weekly Meetings

★ Recognition & Awards

★ Camaraderie & Friends

Teamwork can be a powerful force, but are teams without pitfalls of their own? Are there any disadvantages to working as a team?

Teams Can Be Unwieldy.[54] Anyone who has ever tried to get a group of friends to agree on what kind of pizza to order understands that reaching a consensus and coordinating plans with several people can be maddening.

Teams Pressure Individuals to Conform.[55] Every teen knows that in a group setting, people sometimes tailor their behavior in anticipation of how others will judge it. *Through conformity, the process by which an individual's attitudes, beliefs, and behaviors are influenced by other people, teams can be deprived of the creative thinking and individualism they need to succeed.*[56] Conformity can result in a desire to please others at any cost, to avoid being out of step with others even if one believes their teammates are wrong, to fear being rejected by the group, or to avoid the criticism that follows an unpopular decision.[57]

"Free riders enjoy the fruits of the team's labor without doing their fair share ."

Free Riders. Can a leader be certain that everyone is giving their best? *There's always a chance that there may be free riders, people who receive the fruits of the team's labor without doing their fair share of the work.*[58] Imagine if the fire department relied on donations, like a charity. Some citizens would choose not to contribute, confident that others would provide the firefighters with the resources they need. Team leaders need to be aware of what economists call, "the free rider problem."

Groupthink. One aim of teamwork is to promote group cohesiveness, the ability of the team to stick together and become a united whole.[59] But can cohesiveness go too far? Psychologist Irving Janis believes teams can fall prey to groupthink. *Groupthink occurs when team members seek unanimous agreement in spite of facts pointing to another conclusion.*[60] Put another way, groupthink is a herd mentality. Dumb bulls follow one another to the slaughter. If they know that death awaits, they don't object.

GROUPTHINK
How Teams Kill

On an icy morning in 1986, NASA launched the *Challenger* even though icicles could be seen hanging from the shuttle. Just 73 seconds into its flight, *Challenger* exploded. All seven astronauts perished.

Testifying in Congress after the tragedy, Physicist Richard Feynman used an ordinary clamp and rubber tubing to demonstrate how *Challenger's* O-rings would be affected by launch day's extremely cold weather. "A-ha!" was the senators' and representatives' overall reaction to this simple demonstration. But why didn't NASA see the problem before it was too late?

Some historians believe NASA officials were blinded by "groupthink," which occurs when teams are very cohesive, but insulated from other people and other ideas. If only NASA had invited outside people to help them decide if *Challenger* was a "go" for launch, perhaps the tragedy might have been averted.[61]

Lack of Accountability. As discussed in chapter one, the chain of command is based on the principle that final responsibility for getting a job done ought to be vested in a single individual. Teams, on the other hand, can sometimes represent "leadership by committee." *If everyone on the team is in charge, no one is in charge. If everyone shares responsibility, no one owns responsibility.* Therefore, the leaders who charter teams need to communicate standards and make clear how the team members will be held accountable for their performance.[62]

In leading in a team environment, the leader's challenge is to exploit the team's capabilities. He or she needs to promote teamwork to get the most from its members. On the other hand, the leader also needs to be on guard against the team falling prey to teamwork's many potential pitfalls.[63]

WHO WANTS TO GO FOR ICE CREAM?

A cadet drill team practiced hard every Saturday. For several weeks, they'd end their time together by climbing into the van and driving to the mall for ice cream.

After one practice, C/Capt Earhart, the cadet commander, proclaimed, "Great job everyone, I guess it's time for ice cream." But secretly she had no interest in prolonging an already long day by going for ice cream yet again.

"Yeah, I guess it is time for ice cream," replied her deputy, C/2d Lt Mitchell. With a ton of schoolwork waiting for him at home, he knew he really didn't have time for ice cream this week, but as a good deputy, he wanted to support his cadet commander.

"Well, we better get going. Everyone hop in the van," said the squadron commander, Major Wilson. His wife would have wanted him to come straight home after drill team practice, but "Hey," he said to himself, "I can't disappoint the cadets and say no to ice cream tonight."

So off went the drill team to the mall to get ice cream that no one really wanted. Why?

This story, a retelling of a famous leadership parable known as "The Road to Abilene," is an example of the false consensus.[64] A special kind of groupthink, the false consensus occurs when individuals choose not to express their true feelings about an idea for fear of going against the group.

THE ASCH PARADIGM:
What if everybody says you're wrong?

Which of the three lines on the far right (A, B, or C) is identical to line 1 on its left? What if everyone else tells you you're wrong?

The Asch experiments demonstrated the power of conformity in groups.

When psychologist Solomon Asch asked his simple question about matching lines, there was a twist. He did *not* ask the question to a single individual, but to a group of people.

And, everyone in the group was in on Asch's secret. Everyone except for one individual, who unknowing was the *real* subject of the experiment.

As Asch asked his simple questions, his friends would shout the wrong answer, as they had been secretly instructed beforehand.

When everyone else in the room tells you something you know to be wrong – the sky is green and water is

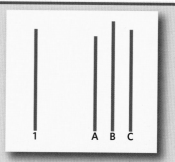

yellow, 2+2=5, or line 1 is the twin of line B – how would you react?

Approximately 75% of Asch's subjects went along with the group. They conformed, even though they knew the group was wrong.[65]

THE TEAM'S LIFE CYCLE

How are teams made? What does it take to transform a bunch of people into a unified team? One researcher, Bruce W. Tuckman, argued that most teams typically pass through four stages on their journey from disorganized group to effective team. This model is so easy to remember that "forming, storming, norming, and performing" has become a popular leadership slogan.[66]

Stage 1. Forming. *When a group first comes together, there is usually chaos.* Everyone is pointed in a different direction. People may not know what to expect. That uncertainty may make people fearful. And having not yet invested time or energy in the team, its members are apt to have a limited commitment to its success. When a bunch of nervous and confused first year cadets report to their flight for encampment, they're forming.

Stage 2. Storming. *As the team begins to take shape, individuals' personalities begin to show themselves.* People struggle to assert their personal needs and goals. Some may battle for attention. As these competing personalities and individual needs clash, team members come into conflict with one another. At this early stage, the team lacks the trust necessary to truly work in unison.

Stage 3. Norming. *Now the team is coming into its own.* The leader's standards gain acceptance by the team and the team members themselves set standards about how the team will work together. Because the uncertainty of the forming stage and the conflict of the storming stage is dying down, people feel more secure. They become more committed to the team's mission and one another. The team is more successful.

Stage 4. Performing. *At last the group has truly become a team. Performing is the stage at which the team is at its best.* After what was probably a rocky start, the team is now entirely focused on the team's goal. Although there's always room for improvement, here the team is fine-tuning its ability to work together. The underlying fundamentals are in place for the team not only to succeed but to reach its full potential.

Team leaders need to be aware of the "forming, storming, norming, and performing" principle of team dynamics so they can provide the right support at the right time. A different style of leadership is needed based on the team's situation. Situational leadership will be discussed in depth in a later chapter.

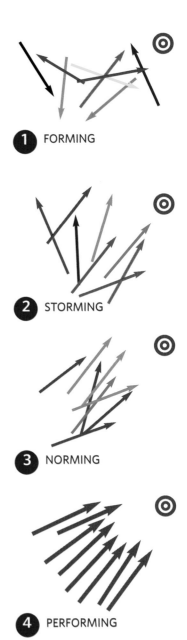

1 FORMING

2 STORMING

3 NORMING

4 PERFORMING

THE L.E.A.D. MODEL

With so much to understand about motivation, coaching, morale, team dynamics, team goals and individual goals, shared purpose, and more, where is a leader to begin? Are there signposts to help a new NCO lead a team?

Yes. *Leadership researchers create models, easy to understand guidelines that boil complex theories down to something workable.* One model is called the L.E.A.D. model. Here's how it works:[67]

Leadership Functions	Leader's Tasks	Team Members' Tasks
Lead with a clear purpose. **L**	Set boundaries	Use active listening
	Explain what the goals are and why	Ask good questions to ensure understanding
	Help the team set its own goals	Participate in setting and clarifying the team's goals
	Evaluate how well the team is reaching the goals	Help leader track the team's progress via feedback
Empower to participate **E**	Ask open-ended questions	Contribute ideas from own experience & knowledge
	Listen actively	Listen actively
	Show understanding	Consider others' ideas
	Summarize what the team is saying	Build on others' ideas
	Seek different views and welcome dissent	Ask open-ended questions
	Record ideas	Think creatively
Aim for consensus **A**	Use brainstorming	Focus on common interests and goals
	Ask questions	Listen to and consider others' ideas
	Listen actively	Make own needs known
	Seek common interests	Disagree in a constructive way
	Summarize	
	Confront and debate ideas in a helpful way	
Direct the team **D**	Give clear directions	Listen actively
	Help the team work together	Keep purpose in mind
	Monitor the team's morale	Stay focused on the mission
	Keep the team focused	Use own energy and enthusiasm to work together
	Encourage people	
	Reward people who go above and beyond	

When new NCOs are beginning to lead the team, they should remember LEAD. It's an easy way to keep the basic principles of team leadership in mind. *By following the LEAD model, the leader will meet all seven needs of a team* that were identified earlier.

FINAL ANALYSIS

In conclusion, the challenge before a new NCO is to transition from a follower to a leader, from a simple team member to one who can influence and unite the whole team. Perhaps the key to successfully making that transition lies in a commitment to that special quality called professionalism. Further, leading a team requires a basic understanding of coaching, constructive discipline, team dynamics, and more.

This chapter only scratched the surface. With so much to learn and absorb, once again we are left with more questions about leadership than we have answers. Regardless, having deepened your perspective on leadership you are now one step closer to your goal of becoming an effective leader.

DRILL & CEREMONIES TRAINING REQUIREMENTS

As part of your study of this chapter, you will be tested on your ability to lead an element in drill and ceremonies. Ask an experienced cadet to help you develop your command voice and practice calling commands on the correct foot. For details, see the *USAF Drill and Ceremonies Manual* available at capmembers.com/drill.

From the Air Force Drill & Ceremonies Manual, Chapter 2

The two main types of commands: the preparatory command and the command of execution

Characteristics of an effective command voice

From the Air Force Drill & Ceremonies Manual, Chapter 3

Command the element to fall in.

Command the element to dress right and check its alignment.

Command the element to perform facing movements.

Command the element to perform flanks and march to-the-rear.

From the Air Force Drill & Ceremonies Manual, Chapter 5c

Manual of the guidon, to include order guidon, carry guidon, rests, salute at the order, and present guidon.

ENDNOTES

1. Samuel P. Huntington, *The Solider and the State*, (Cambridge, Mass: Harvard University Press, 1957), 17.

2. Civil Air Patrol, *Leadership for the 21st Century*, (Maxwell AFB: CAP, 2004), 34.

3. Daniel Goleman, *Primal Leadership*, (Boston: Harvard, 2002), 173.

4. U.S. Army, AFM 7-22.7, *The Army Noncommissioned Officer Guide*, (U.S. Army, 2002), 1-32.

5. Peter Drucker quoted in *Heirpower!* by Bob Vasquez, (Maxwell AFB: Air University Press), xvii.

6. 42d Air Base Wing, Maxwell AFB, Al, Top Three Council.

7. U.S. Air Force, AFI 36-2618, *The Enlisted Force Structure*, (U.S. Air Force, 2009), ch 4.

8. Ibid.

9. "The Chief Master Sergeant of the U.S. Air Force," (U.S. Air Force, 2010), http://www.af.mil/information/bios/bio.asp?bioID=12543.

10. James A. Autry, *The Servant Leader*, (New York: Three Rivers Press, 2004), 20.

11. Robert K. Greenleaf, *The Servant as Leader*, (Indianapolis: Greenleaf Center, 1994 ed.).

12. Mark 10:42-44, NRSV.

13. Martin Luther King Jr., quoted at The King Center, thekingcenter.org, Jan 2010.

14. Autry, 21.

15. Rodney J. McKinley, "Caring is Free," CMSAF Viewpoints, (U.S. Air Force, Apr 2008), http://www.af.mil

16. Retold from "Man Enough for the Job," in *The Moral Compass*, William J. Bennett, ed. (New York: Simon and Schuster, 1995), 657.

17. *Harvard Business Essentials, Coaching & Mentoring*, (Boston: Harvard, 2004), 2.

18. Ibid, 2.

19. Brian Emerson & Anne Loehr, *A Manager's Guide to Coaching*, (New York: American Management Association, 2008), 22.

20. Ibid, 20-23.

21. Ibid, 86-87.

22. Louis V. Imundo, *The Effective Supervisor's Handbook*, 2d ed., (New York: American Management Association, 1991), 4-10.

23. Paul Hersey, Kenneth H. Blanchard, & Dewey E. Johnson, *Management of Organizational Behavior*, 7th ed., (New York: Prentice Hall, 1996), 287.

24. Ibid, 278.

25. Ibid, 284-287.

26. Moses Simuyemba, "Definition of Motivation," Jan 2010, http://www.motivation-for-dreamers.com/definition-of-motivation.html.

27. John C. Maxwell, *The 17 Essential Qualities of a Team Player*, (Nashville: Nelson Business, 2002), 23.

28. Goleman, 61.

29. Changing Minds.org, Jan 2010, http://changingminds.org/explanations/theories/intrinsic_motivation.htm.

30. Kathy MacKay, "The Sultan of Svelte," *People*, Nov 2 1981, 94-95.

31. Robert Heller, *Learning to Lead*, (New York: DK Publishing, 1999), 58.

32. Thomas Wolf, *Managing a Nonprofit Organization*, (New York: Prentice Hall), 90.

33. Steve Cox, "Stone, CAP Charter Member, Dies," *CAP Volunteer Now*, Sept 15, 2009.

34. Civil Air Patrol, 43.

35. U.S. Air Force, AFDD 1-1, *Leadership & Force Development*, (U.S. Air Force, 2006), 9.

36. Ibid, 16-18.

37. Ibid, 9-10.

38. Michael Hieb & Ulrich Schade, "Formalizing Command Intent...," *The International C2 Journal*, vol. 1, no. 2 at http://www.dodccrp.org/html4/journal_v1n2_04.html.

39. *Oxford American Dictionary*.

40. William Timothy O'Connell, "Military Dissent and Junior Officers," in AU-24 *Concepts for Air Force Leadership*, (Air University Press: 2001), 324-326.

41. Philip S. Meilinger, "The Ten Rules of Good Followership," in AU-24, *Concepts for Air Force Leadership*, (Air University Press: 2001), 99-101.

42. The Churchill Centre, "Churchill and War," retrieved Jan 2010 at http://www.winstonchurchill.org/learn/biography/the-admiralty/churchill-andwar

43. Stephen R. Covey, *Principle-Centered Leadership*, (New York: Free Press, 1991), 238.

44. Ibid, 238.

45. Meilinger, 101.

46. The Ken Blanchard Companies, "The Critical Role of Teams," retrieved Jan 2010 at http://www.kenblanchard.com/Business_Leadership/Leadership_Theory/critical_role_teams.

47. Fran Rees, *How to LEAD Work Teams*, (San Diego: Pfeiffer & Company, 1991), 38-41.

48. Mitch McCrimmon, "What is Informal Leadership?," (suite101.com, July 2007), Jan 2010.

49. *Oxford American Dictionary*.

50. Ibid.

51. The Ken Blanchard Companies.

52. Jean-Paul Sartre, *Huis Clos* (No Exit), Act I, Scene 5.

53. Hal G. Rainey, *Understanding & Managing Public Organizations*, 3rd ed., (San Francisco: Jossey-Bass, 2001), 341.

54. Ibid, 336.

55. Ibid, 334.

56. *Oxford American Dictionary*.

57. Harvard Business Essentials, *Decision Making*, (Boston: Harvard, 2006), 117.

58. *Stanford Encyclopedia of Philosophy*, "Methodological Individualism," March 2009, http://plato.stanford.edu/entries/methodological-individualism/, Jan 2010.

59. *Oxford American Dictionary* ("cohesion").

60. Irving L. Janis, "Groupthink," in *Classics of Organization Theory*, (Orlando: Harcourt Brace, 1996), 184.

61. Edward Tufte, *Visual Explanations*, (Cheshire, Conn: Graphics Press, 1997), 50-51.

62. The Ken Blanchard Companies.

63. Rainey, 334.

64. Jerry B. Harvey, "The Abilene Paradox..." in *Organizational Dynamics*, Summer 1974.

65. Morton Hunt, *The Story of Psychology*, (New York: Anchor, 2007), 460-61.

66. Bruce W. Tuckman, "Developmental Sequence in Small Groups," in *Psychological Bulletin*, no. 63, 384-399.

67. Rees, 54.

PHOTO CREDITS

CHAPTER 5
BRAINPOWER *for* LEADERSHIP

LEADERSHIP IS AN INTELLECTUAL ACTIVITY. It requires brainpower. Developing your brainpower can only make you a more effective leader.

Recognizing this fact, the U.S. military, the most professional force in the world, requires all officers to be college graduates. Moreover, even after commissioning, officers will attend four or more graduate-level schools on their way to the grade of colonel and above. Top NCOs receive a rigorous education, too.

Moreover, a leader's overall brainpower will affect his or her success. In this chapter, we look at critical thinking, creativity, and teaching – three aspects of leadership that speak to the need for serious study and the application of brainpower.

Leaders need to teach themselves how to think and how to learn. If you want to lead, you better get yourself smart.

CRITICAL THINKING

"To every complex question there is a simple answer, and it is wrong."

H.L. MENCKEN
Journalist and social critic

OBJECTIVE:

1. Defend the claim that critical thinking has a direct impact on a leader's effectiveness.

If leadership requires careful study and reflection, as we discussed in chapter three, then *a leader's critical thinking skills have a direct influence on his or her effectiveness.* The good news is that everyone can develop better critical thinking skills through study and practice.

After all, great ideas rarely go out and find someone. It's up to leaders to exercise their brainpower better than their competitors. Lazy thinkers get left behind. In contrast, as leaders become better at thinking critically, their rewards grow. They imagine better ideas and solve problems more quickly.

The World War I "Ace" Eddie Rickenbacker had the right idea when he said, "I can give you a six-word formula for success: think things through, then follow through."[1]

CHAPTER GOALS

1. Explain why and how leaders try to be critical thinkers.

2. Appreciate the value of creativity in leadership.

3. Develop an understanding of some fundamentals in teaching and training.

CHAPTER OUTLINE
In this chapter you will learn about:

Critical Thinking
  Universal Intellectual Standards
  Modes of Thinking
  Logical Fallacies

Creative Thinking
  Unappreciated Geniuses
  Monuments to the Status Quo
  Victories Through Creativity
  Tools for Creative Thinking

Teaching & Training People
  The Trainer's Goalposts
  Learning Styles
  Teaching & Training Methods
  Evaluating Learning

Drill & Ceremonies

PRINCIPLES OF CRITICAL THINKING

OBJECTIVE:

2.   Define the term, "critical thinking."

Critical thinking is self-guided, self-disciplined thinking which attempts to reason at the highest level of quality in a fair-minded way.[2] Put another way, critical thinkers value reason and they work hard to avoid letting their own prejudices, assumptions, or emotions cloud their logic. But don't be confused by the word "critical" because in this context it does not mean to nitpick someone else's idea or always try to find something wrong just to prove the other person isn't perfect. Rather, ***critical thinking is the habit of being guided by universal values of logic and a deep respect for the truth.***

As with other aspects of leadership, becoming a critical thinker is more a journey than a destination. Everyone is subject to lazy thinking or irrational thought from time to time. Therefore, ***developing the ability to think critically is a lifelong endeavor, a never-ending process.*** The great philosopher Socrates expressed this idea when he said, "The unexamined life is not worth living."[3] Taken together, many unexamined lives result in an uncritical, unjust, and dangerous world.[4]

> **"Critical thinkers try to prevent prejudices and emotions from clouding their logic."**

"[Critical thinking] is a desire to seek, patience to doubt, fondness to meditate, slowness to assert, readiness to consider, carefulness to dispose and set in order; and hatred of every kind of imposture."

SIR FRANCIS BACON
One of the first thinkers to use the scientific method

SOCRATES: THE THINKER AS HERO

"The unexamined life is not worth living."

The ancient Greek philosopher Socrates fiercely advocated that every person use their brainpower to seek truth, consider what is good, and promote justice. For his teachings, he was sentenced to death.

In the painting, "The Death of Socrates," we see Socrates heroically asserting that no evil can ever truly harm a moral person, although his followers scream and wail in anguish.

The painting's harmony, simplicity, and proportion reflect qualities that Socrates himself found virtuous. Here is a portrait of a man who valued brainpower.

"The Death of Socrates" shows why so many celebrate Socrates as a hero of critical thinking.[5]

UNIVERSAL INTELLECTUAL STANDARDS

OBJECTIVE:

3.   Describe the seven universal intellectual standards.

What are the universal values of logic that all leaders must respect if they are to be considered critical thinkers? Here are seven:[6]

Clarity. "What exactly do you mean by that point? Can you say that again in a different way to help me better understand?" ***The principle of clarity calls for critical thinkers to express their ideas such that people will know exactly what thoughts are racing through their brains.*** If we cannot clearly understand what someone is saying, we cannot honestly evaluate whether their point makes sense.

Accuracy. "Every Spaatz cadet is guaranteed admission into the U.S. Air Force Academy." That sounds too good to be true. ***The principle of accuracy demands that critical thinkers back up their claims and that other people be allowed to double-check those claims.*** Someone might challenge the accuracy of that claim by noting, "The Academy's website doesn't promise Spaatz cadets admission, and there's nothing in the cadet regulations to support that claim."

Precision. "Give or take a million bucks, most cadets are millionaires." This statement tells us nothing because it lacks precision. On the other hand, "a survey of 20 of the 28 cadets in our squadron reveals that the average cadet is 15.2 years old," is a fairly precise statement ***Precise statements mean what they say and say what they mean.***

Relevance. A statement may be clear, accurate, and precise, but not relevant to the issue. ***The principle of relevance calls for all supporting claims to advance the overall argument.*** "Cadet NCOs often hold leadership positions, and the capitol of Idaho is Boise." Because those two points are not related, one of them cannot be relevant to the overall topic. Emotional pleas are often irrelevant. A cadet may want to be promoted very, very badly, but wanting something is different from deserving it. The want is irrelevant.

Depth. Anyone who has listened to classmate deliver a report on a book they didn't read knows about depth. Such a report will be superficial, barely skimming the surface and totally ignoring the main issues. In contrast, ***good critical thinking is marked by depth, the willingness to examine every imaginable complexity or factor bearing on an issue.*** Only someone who has closely read a novel can discuss it in depth.

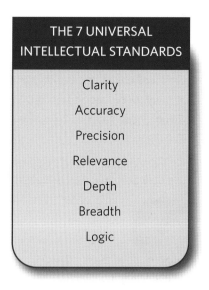

THE 7 UNIVERSAL INTELLECTUAL STANDARDS

Clarity

Accuracy

Precision

Relevance

Depth

Breadth

Logic

Precision.
So much is riding on this cadet's precision with a compass. If his bearing is imprecise, he will be off course down range. The same principle applies to critical thinking. Imprecise arguments can lead you way off target.

Breadth. Depth concerns how deeply a critical thinker is willing to dig into an issue. *Breadth, on the other hand, concerns how far across either side he or she is willing to look when considering an issue.* For example, one could talk about summer encampment in great detail, describing what happens there on a minute-by-minute basis, but if the overall question is "What's the Cadet Program all about?" their argument will lack breadth, having ignored other aspects of cadet life.

Logic. Finally, critically thinking is supposed to be logical. "Does this really make sense? First you said cadets have to be 12 to join CAP, and now you say some cadets are ten?" *When one point supports the next and the conclusions flow naturally, an argument is logical.* If one point contradicts another or the argument doesn't "make sense," the argument is illogical. (We'll return to logic later in this chapter.)

ELEMENTS OF THOUGHT

OBJECTIVE:

4.   Describe the eight elements of thought.

Eight basic structures are present in all thinking. Critical thinking generates purpose, raises questions, uses information, utilizes concepts, makes inferences, makes assumptions, generates implications, and embodies a point of view. The checklist below can be helpful as you work to develop good habits of critical thinking.[9]

1. Reasoning has a purpose.
★ What's your purpose?
★ Can you state it clearly?

2. Reasoning is an attempt to figure something out.
★ Precisely state the question.
★ Express the question in several ways to clarify its meaning and scope.
★ Break the question into sub-questions.

3. Reasoning is based on assumptions or beliefs you take for granted.
★ What assumptions are you making? Are they justifiable?
★ How might your assumptions be shaping your point of view?

4. Reasoning has a point of view.
★ What is your particular point of view?
★ How might your point of view influence how you see a problem?

5. Reasoning is based on data, information, and evidence.

★ Are your claims backed up by data?
★ Have you searched for information that contradicts your assumptions?
★ Is the data trustworthy and relevant?

6. Reasoning is expressed through and shaped by concepts and ideas.
★ Identify key concepts and express them clearly.
★ Consider alternative concepts or alternative definitions of your concepts.

7. Reasoning contains inferences by which we draw conclusions.
★ Are you inferring only what the evidence implies?
★ Do all your inferences point to the same or different conclusions?

8. Reasoning leads somewhere and has consequences.
★ What does your answer really mean? So what?
★ What surprises might result from your ideas?

---

truth•i•ness n.

Comedian Stephen Colbert famously coined the word "truthiness" to describe the squishy quality of truth in a leader's thinking.[7] Truthiness comes from the gut, not the brain, and is supposedly exempt from the rules of logic.

Colbert explains: "What I say is right and [nothing] anyone else says could possibly be true. It's not only that I *feel* it to be true, but that *I* feel it to be true. There's not only an emotional quality, but there's a selfish quality, too."[8]

The zany feeling of truthiness is what results when leaders forget universal intellectual standards.

IMPLICATIONS FOR LEADERS

What do the principles of critical thinking mean for leaders? To become a stronger critical thinker, expose yourself to other good thinkers. Read a serious newspaper, a challenging novel, or a work of non-fiction that has something important to say. Hang around the smartest people you know and ask them lots of questions. Sign up for challenging math courses to stretch your mental muscles. Become a leader who has a habit of thinking sharply.

MODES OF THINKING

OBJECTIVE:

5.   Explain four different modes of thinking.

Leaders' critical thinking skills take on several different forms. Call these the modes of critical thinking. Some of these modes include big-picture thinking, focused thinking, realistic thinking, shared thinking, and creative thinking. A summary of each is included below (though creative thinking will be discussed later).

BIG-PICTURE THINKING

"You can find many big-picture thinkers who aren't leaders," reports one expert, "but you will find few leaders who are not big-picture thinkers."[11] Put simply, ***big-picture thinking is the practice of stepping back from an issue or problem so as to take more of it in.*** Big-picture thinkers see the full breadth of the situation. Philosopher and Roman Emperor Marcus Aurelius showed his respect for big-picture thinking when he wrote, "Look always at the whole."[12]

"You've got to think about 'big things' while you're doing small things," says one leadership theorist, "so that all the small things go in the right direction."[13] ***Big-picture thinking helps leaders stay on target. Further, it promotes teamwork.*** When looking at the big picture, it's only natural that you will notice how the various members of a team support one another and help fulfill the mission. Finally, ***big-picture thinkers are able to synthesize or mesh together their learning.*** Instead of locking away every individual thing they learn into its own drawer, big-picture thinkers look for ways to synthesize their learning. An average student may do okay in two different subjects at school – history and English, for example – but the big-picture thinkers become outstanding students because they see how what they learn in history connects with what they study in English, or what they learn in physics relates to what they learn in algebra.

HOW FACTS CHANGE EVERYTHING
(If You Let Them)

How can leaders present their critical thinking in an age where PowerPoint's natural "selling" style of presentations has corrupted our ability to consider problems thoughtfully?[10]

★   Concentrate on delivering facts, not cute slogans or unsupported claims.

★   Deliver as many of those facts as you can.

★   Respect your audience's intelligence.

★   Look to serious journalism for examples of solid reporting and analysis. Good journalism, like rigorous thinking, is grounded in facts and describes how one event caused the next.

★   Use good teaching as your metaphor when presenting a proposal. Try to "teach" your audience why your idea works.

★   Remember that PowerPoint does not persuade, only great critical thinking persuades.

If you explain your idea clearly and support your claims, your argument should sell itself.

FOCUSED THINKING

The tougher the problem, the harder you have to think about it. "No problem," according to the writer Voltaire, "can withstand the assault of sustained thinking."[14] *Focused thinking is the practice of intensely studying an issue, trying to see it clearly, and not becoming distracted by other issues that are somewhat related to, but different from, the specific question at hand.* Focused thinking is more about deep concentration than unlimited imagination. "The immature mind hops from one thing to another," according to one philosopher who clearly appreciated focused thinking; "the mature mind seeks to follow through."[15]

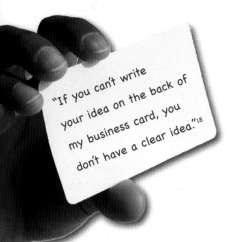

"If you can't write your idea on the back of my business card, you don't have a clear idea."[18]

One way to put focused thinking into practice is by using the 80/20 rule. That is, devote 80% of your time and energy to the top 20% of the issues you face.[16] Spending your brainpower on the biggest challenges you face should yield the biggest results. Moreover, the principles of focused thinking run counter to how many youth live their lives in the 21st century. Instead of switching from task to task all the time – watching TV while chatting online, while writing a term paper, while petting your dog – focused thinking demands you put every ounce of your brainpower toward one single issue. *Researchers have found focused thinking is more efficient because it allows the brain to work better, while unfocused thinking can actually lower a person's IQ.*[17]

> "There's nothing like staring reality in the face to make you see the need for change."

REALISTIC THINKING

If you are the type of leader who is naturally optimistic, perhaps you sometimes need to be brought back to the real world. *Realistic thinking is an approach where the leader tries to see the world for how it actually is, not how we might wish it to be.*[19] Accuracy, common sense, and feasibility are some of the key values in realistic thinking. One realist put it this way: "There's nothing like staring reality in the face to make a person recognize the need for change."[20]

Another way to understand realistic thinking is to consider the leader's never-ending need to balance shortfalls. Air Force General Stephen Lorenz explains:

> Shortfalls occur in our professional and personal lives. We never seem to have enough time, money, or manpower. The essence of this "scarcity principle" lies in accepting the reality of limited resources and becoming adept at obtaining superior results in less-than-ideal situations. Equally important, once people acknowledge the scarcity of resources, then they

The BUCK STOPS *here!*

Realistic & Responsible.
"I never give 'em hell. I just tell the truth and they think it's hell."[21]
PRESIDENT HARRY S. TRUMAN

Are you a dreamer or a realist? Dreamers imagine the world as it could be. Realists insist on seeing the world as it is. Which is the better approach for a leader? Can you be both?

need not bemoan the situation any longer. In other words, they should "deal with it." Leaders must carry out the mission with the resources they have. They have to make it happen![22]

To become better at thinking realistically, begin by getting all the facts. Are they accurate and relevant? Then picture the worst case scenario and use realistic thinking to prepare for it. Seek out great minds and ask for their help. *Leaders who are realists gain credibility because they operate in the real world and are prepared for whatever curve balls come their way.*[23]

Humble Leaders Use Shared Thinking. President Ronald Reagan must have understood the need for emotional security when he said, "There is no limit to what you can accomplish if you don't care who gets the credit."[30]

SHARED THINKING

Shared thinking involves valuing the thoughts and ideas of others.[24] It is a mode of thinking that comes from an appreciation for team-work and the belief in synergy. One of its principles is that all of us working together are smarter than any one of us working alone. Shared thinking is faster than other forms of critical thinking, too. When you need to understand a complex issue, it's usually quicker to ask a trusted expert than to go research the issue from scratch.[25]

Synergy.
The belief that a team is greater than the sum of its parts. See chapter 2.

In chapter one, we discussed the shortcomings of "leadership by committee." At first glance, shared thinking seems to have similar drawbacks. But in 1787, fifty-five men demonstrated the greatest example of shared thinking as they worked together to craft the Constitution of the United States. Instead of a constitution that served only individual interests, shared thinking made for a lasting republic.[26]

Is shared thinking easy to do? *The biggest obstacle to shared thinking is emotional insecurity.*[27] One expert explains, "People who lack confidence and worry about their status, position, or power tend to reject the ideas of others, protect their turf, and keep people at bay. It takes a secure person to accept another's ideas."[28]

Shared Thinking Works.
Fifty-five leaders, working together, produced the greatest political document in world history: The Constitution of the United States of America.

LEADERS THINK CRITICALLY

Universal intellectual standards, the elements of thought, and modes of thinking help explain what critical thinking is and how leaders practice it. What we know about critical thinking reinforces the principle that the work of the leader includes maximizing his or her brainpower. Those who fail to realize this, or those who prefer lazy thinking, make the madman's dreams come true. As Hitler wrote, "What luck for rulers that men do not think!"[29]

Make the Nazis Suffer. Hitler once exclaimed, "What luck for rulers that men do not think!" If the worst thugs in human history hated it, critical thinking must be an incredible force for freedom.

THE TWISTED THINKING OF LOGICAL FALLACIES

OBJECTIVES:

6.   Explain what logical fallacies are and how they affect leaders.

7.   Give examples of at least five different types of logical fallacies.

A logical fallacy is an error of reasoning.[31] When someone makes an argument based on bad reasoning, they are said to commit a fallacy. Weak, twisted, fallacious thinking keeps us from knowing the truth. Therefore, to be good critical thinkers, leaders must study logical fallacies, both so they can avoid using them and spot them in others.

Fallacies are so common (even among the brightest minds) that philosophers have been able to define these recurring mistakes and give each a name. Here are ten of the most common logical fallacies:[32]

Ad Hominem (Latin, literally "to the man"). Have you ever see someone who is losing an argument make a personal attack on their opponent? That's an *ad hominem* attack. ***Instead of focusing on the logic of an opposing argument, an*** **ad hominem** ***attacks the other person.*** In addition to being logical fallacies, *ad hominem* are simply rude.

EXAMPLES:

• "What can our new math teacher know? Have you seen how fat she is?"

• "Why would I listen to that moron?"

• "So I'm ugly. So what? I never seen anyone hit with his face." (Baseball great Yogi Berra, in response to an *ad hominem*)[33]

Appeal to Authority. Perhaps the weakest of all fallacies, ***an appeal to authority tries to prove a claim by asserting that some smart person believes the claim to be true and therefore it must be true.*** Anyone who has ever spent time on the playground has seen kids try to win arguments with the devastating, "Ya-huh! My big brother said so!"

EXAMPLES:

• "The purpose of the Cadet Program is not to recruit for the Air Force."
REPLY: "Yes it is. Cadet Curry said so and he outranks you."

• "I admire the president for being a good role model."
REPLY: "But the *New York Times* says he's a lousy role model, so you're wrong."

Post Hoc Fallacy. *Post hoc ergo propter hoc* (Latin: After this, therefore because of this). Something happens and then something else happens. Does that mean that the first thing caused the second? Not necessarily. ***The post hoc fallacy illustrates the difference between correlation (two things being related somehow) and causation (one event causing another event, like a chain reaction).***

EXAMPLES:

• Shortly after MySpace became popular, U.S. soldiers found Saddam Hussein.

• Michael Jackson, Kurt Cobain, and Jimi Hendrix were rock stars who died young. Therefore, if you become a rock star, don't expect to live a long life.

Fallacy.
A mistake in logic; bad reasoning that corrupts a line of thought.

Note: These explanations have been simplified, but are accurate enough for our purposes.

THINK

It is one of leadership's all-time greatest slogans: THINK.

Thomas J. Watson, the first president of computer giant IBM, coined the powerfully simple motto a century ago. He reasoned:

"Thought has been the father of every advance since time began. 'I didn't think' has cost the world millions of dollars."[33]

You'll find THINK displayed in IBM offices and factories, on their websites and annual reports and everywhere in between.

Watson valued clear thinking and innovation. How would he get his team to understand his vision? THINK.

Appeal to Tradition. It's the "We've always done it that way" response. One of the quickest ways to lose credibility as a leader is to commit the fallacy of the appeal to tradition. ***This fallacy makes the assumption that older ideas are better, and that the leader's job is to prevent change.***

EXAMPLES:

• "If we allow cadets to apply for encampment online, we'd save everyone lots of headaches."
REPLY: "No. We've always made cadets apply using a paper form."

• "We should offer movies on our company's website."
REPLY: "No, we've built our company's fortune by renting movies only through our stores."

Red Herring. One of the most seductive fallacies, the red herring is a distraction. ***And while a given line of thought may indeed be true, it is a red herring if it is not relevant to the issue at hand.*** Their truthfulness makes red herrings particular effective at derailing someone's successful argument.

EXAMPLES:

• "We should present Cadet Curry with our squadron's cadet of the year award. He's the most active and highest-performing cadet we have."
REPLY: But Cadet Arnold has been in CAP longer.

• "I know you want to imprison me for having murdered my parents, but judge, have mercy on me, I'm an orphan!"

Weak Analogy. It's "apples and oranges." People often make analogies or comparisons. They see how one situation or one thing is similar to another, and indeed they are. ***But the fallacy of the weak analogy arises because no matter how similar two things are, they are never exactly alike,*** and therefore, the argument breaks down.

EXAMPLES:

• Hybrid cars are like solar power, full of promise but too expensive. We'll never be able to build affordable hybrid cars.

• Encampment is like basic training. It's CAP's equivalent to Boot Camp.

Straw Man Fallacy. The most persuasive debaters fearlessly attack the opposing argument in its *strongest* form. To truly show their solution is superior, they try to demonstrate that the opposition cannot compete even on its best day. In contrast, some shrink from a fair quest for the truth by setting up a straw man argument. ***Instead of attacking the opposition head-on, a straw man fallacy misrepresents the opposing position, making it seem weaker than it is.***

EXAMPLES:

• We should prohibit all cadets from assisting on emergency services missions. Some cadets are so immature they won't ever stop goofing around.

• Senator Curry is against the new F-99 laser fighter. I'm for it because I don't want to leave America defenseless.

DO LITERAL THINKERS BAKE YUMMY CAKES?

Precision, clarity, accuracy. Logic asks us to focus intently on what is being said. But maybe we shouldn't focus *too* closely, like these cake decorators did.

Begging the Question / Circular Reasoning. *If your argument's conclusion is the same as one of your premises, you're begging the question.* Your reasoning is running in circles. The only people likely to be persuaded by circular reasoning are those who already agree with the original premise. When someone tries to support a statement by restating it again and again, they are said to be begging the question or using circular reasoning.

EXAMPLES:

- "You can't give me a C. I'm an A student!"
- "Honesty is defined as always being honest."
- "My door must have been locked. I always lock my door."

False Dilemma. *The premise behind the false dilemma is that we are faced with two, but only two choices, and both are not very good.* It's an "either / or" situation; you can have either X or Y, but not both, and you certainly cannot have Z. This fallacy often arises because of the inability to think creativly and see an acceptable third way to a solution.

EXAMPLES:

- "Are you going to do well in school, or are you going to succeed as a cadet?"
- "We can give cadets awards for doing well, or we can have a disciplined squadron. We can't do both."

"False dilemmas arise because of a failure to think creatively."

HELL WAS IN SESSION, BUT HE HAD OTHER PLANS

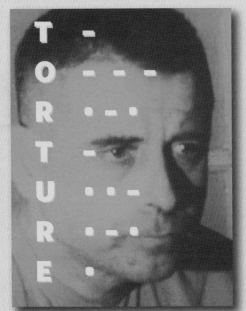

Rear Adm. Jeremiah Denton USN
Recipient of the Navy Cross
Author of *When Hell Was In Session*

"You're going to betray America on TV, or we're going to kill you."

Can an honorable military officer encounter a worse dilemma? If a dilemma is the choice between two equally disastrous options, Commander Jerry Denton, USN, sure was facing a dilemma while held as a prisoner of war in the infamous Hanoi Hilton during the Vietnam War.

What would he do? What would you do?

As a highly-educated officer, CDR Denton knew something about critical thinking and the fallacy of the false dilemma. He knew that he had to reject the two choices being forced upon him by his captors, but how?

Denton recalled, "My only firm conviction was that I would die of starvation before I would [make] a confession."

As he was being pushed in front of TV cameras, his eyes blinked under the bright lights. That was it! He found a third way, a way out of his false dilemma. Denton kept blinking his eyes rhythmically, but with a purpose. In Morse code he spelled "t-o-r-t-u-r-e, t-o-r-t-u-r-e, t-o-r-t-u-r-e."[34]

It was the first clear message that American officials received that the POWs were being tortured. Through his ingenuity, CDR Jerry Denton retained his honor *and* found a way to tell America what was really going on in the Hanoi Hilton.

Slippery Slope. *If the idea that one thing leads to another is fixed in your mind, you may travel down the slippery slope.* Pull the trigger, and the gun fires. Indeed, one thing is apt to cause another, but in critical thinking, causation must be shown, it can't be assumed.

EXAMPLES:

• Major in English in college, start reading poetry, and next thing you know you'll become an unemployed pot-smoking loser.

• Never be kind and generous to the poor. They'll come to expect your help always and never learn to contribute to society.

INTEGRITY & THE CRITICAL THINKER

OBJECTIVE:
8. Discuss the concept of intellectual honesty.

Why do many of the fallacies described above seem so familiar? Despite fallacies being examples of twisted thinking, sometimes they sneak past us and we get duped by them. That fact shows the need for every leader (or every responsible citizen for that matter) to be familiar with the most common fallacies. *Those who search for the truth need to be on guard against logical fallacies, both in their own arguments and in the arguments of others.*

These principles of integrity comprise something called *intellectual honesty – honesty in the acquisition, analysis, and transmission of ideas.*[38] Or to put it more simply, intellectual honesty means you exercise your brainpower with integrity. One professor's view on intellectual honesty is worthy quoting at length:

> …Human beings are more than mere purveyors of logic. We inherently generalize, categorize, prioritize and harmonize what we see, and most of this takes place without [us realizing what we're doing]. While these aspects of thinking are of valuable, they also possess certain dangers. For example, they can lead us into hasty judgments, and cause selective "blindness" toward new information. Intellectual honesty is one [way to watch out for those] prejudices, by forcing us to examine how we arrived at them… Once we become aware of these pitfalls in thinking, it then becomes a matter of choice as to whether we attempt to compensate for them.[39]

In other words, even *how* we think has moral consequences. The job of the leader just got that much tougher.

Try Lifting This.
If anyone speaks of God, whether they profess a certain faith or not, they're speaking of an all-powerful being. There is nothing this God can't do, right? Well, if he's so powerful, can he make a rock that's so humongous even he can't lift it? This tricky old question is a *paradox*, a line of reasoning that points to an impossible answer.

RUSSIAN SELF-DECEPTION

If you turn a blind eye to critical thinking, what happens?

Can you be so good at lying to yourself that you begin to believe your own lies? The great Russian novelist Fyodor Dostoevsky thought so.

from
THE BROTHERS KARAMAZOV:

"Above all, do not lie to yourself. A man who lies to himself and listens to his own lie comes to a point where he does not discern any truth either in himself or anywhere around him, and thus falls into disrespect towards himself and others."[40]

49

CREATIVE THINKING

OBJECTIVE:

9.   Explain what "creative thinking" means in your own words.

Creative thinking is concentration plus imagination. It is the habit of trying to see ideas or objects in a new context. Creative thinking is an attempt to grab hold of an invisible thread connecting two concepts. It requires us to overcome how we are constrained by culture, tradition, or circumstance. But for creativity to be meaningful, it must produce results. One expert contends that creativity must produce "work that is both novel and appropriate."[41] Therefore, a creative leader will have contempt for the "we've always done it that way" attitude. **In CAP, our Core Value of Excellence requires us to think creatively.**

Note: Throughout this section, we focus on creative thinking and how creativity impacts leadership. We won't consider creativity from an artistic point of view.

Some outstanding creative thinkers have this to add:

"A foolish consistency is the hobgoblin of little minds . . . Speak what you think now in hard words, and tomorrow speak what tomorrow thinks in hard words again, though it contradict everything you said today. – 'Ah, so you shall be sure to be misunderstood.' – is it so bad, then, to be misunderstood? Pythagoras was misunderstood, and Socrates, and Jesus, and Luther, and Copernicus, and Galileo, and Newton. To be great is to be misunderstood."[42]

RALPH WALDO EMERSON

"The reasonable man adapts himself to the world; the unreasonable one persists to adapt the world to himself. Therefore all progress depends on the unreasonable man."[43]

GEORGE BERNARD SHAW

"Think left and think right and think low and think high. Oh, the thinks you can think up if only you try!"[44]

DR. SEUSS, THEODOR GEISEL

"There's a party in my mind and I hope it never stops."[45]

TALKING HEADS

COME ON IN, WE'RE PLANNING THE AIR CAMPAIGN

It was a military strategist's dream come true. The phone rings and the general who will lead the Persian Gulf War is on the other end. Col John Warden was being asked to develop an air campaign for the coming fight, but he'd have only two days to do the job. What would you do? Lock yourself into a room to limit distractions and get to work? Not John Warden. He realized the need for open planning:

"I briefly considered gathering a very small group of people around me, closing the doors, and doing it all in great secrecy. Quickly, however, I realized that this didn't make much sense – I was certainly no expert on Iraq, I needed a lot of help, but I didn't even have a way of knowing who or what I needed.

I decided to open the doors of a very big briefing room we had in the basement of the Pentagon and gather as many people as possible. Right from the start, everyone in the group was involved in almost everything that took place. This way, everybody understood not only the decision, but also the thinking and

discussions associated with them. So they were able to do most of their work without reference to any higher authority, secure in the knowledge they were doing the right thing."[46]

Because of Warden's success in envisioning and planning the Gulf War air campaign, he's been called one of the leading air power theorists in Air Force history.

UNAPPRECIATED GENIUSES

OBJECTIVE:

10. Discuss why creative leaders do not always win.

Even if you develop the habit of thinking creatively, there is no guarantee that your ideas will be welcomed. People often resist change, and the status quo is a comfortable place in which to live. The experiences of Billy Mitchell, Galileo, Emily Dickinson, Martin Luther King, and others illustrate this point. Again we return to Emerson: "To be great is to be misunderstood." Here are two examples of failed genius from the world of business:

Status Quo.
The existing state of affairs; the way something's always been done.

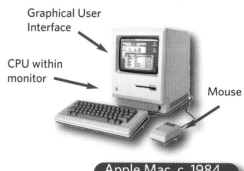

Apple Mac, c. 1984

Apple vs. IBM. In the 1980s and early days of the home computer, Apple was battling IBM for control of the market. The experts agreed that Apple's product was superior. After all, it was the first to use a mouse, the first to offer a friendly interface like today's computers (to run the IBM, you had to physically type complex commands), and Apple's graphics were clearly better. Objectively speaking, Apple offered a superior product. But did that matter? No. By allowing other companies to make IBM-compatible computers (Dell, Compaq, HP, Gateway, etc.), the IBM system quickly dominated the market. Apple was lucky to survive.[47]

IBM PC, c. 1983

Tucker vs. Corvair. In 1948, inventor Preston Tucker envisioned a new kind of car. Disc brakes, fuel injection, and overhead valves made the Tucker years ahead of the competition. If your Tucker broke down, no problem. Your mechanic would simply swap-out the engine and lend you a new motor. Directional headlights, a pop-out windshield and a padded dashboard were some other wonders that made the Tucker obviously superior to the cars of its day. Despite its awesome features, Tucker sold only 51 cars in 1948.[48]

The Chevrolet Corvair of 1960-65 was perhaps the Tucker's evil twin. One safety critic famously labeled it "unsafe at any speed."[49] The Corvair was prone to roll-overs. Maintaining tire pressure was crucial. Drive too long and you might die from carbon monoxide poisoning. Even the battery had problems – sometimes it emitted hydrogen! Despite its terrible safety and reliability record, Chevy sold over one million Corvairs.[50]

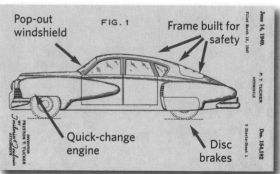

Tucker, 1948

Lesson for Leaders. What can leaders learn from the stories of Apple vs. IBM, Tucker and Corvair? It's tough to be innovative, to think creatively. Even more, ***the most creative leader isn't guaranteed to win***. Wonderful ideas are nothing without strong execution. Creativity, then, involves learning from mistakes.

MONUMENTS TO THE STATUS QUO

OBJECTIVE:
11. Give examples of how majorities can discourage creativity.

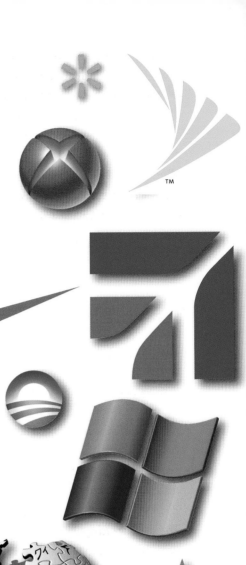

"I know of no country in which there is so little independence of mind and real freedom of discussion as in America."[51] – ALEXIS DE TOCQUEVILLE

What a puzzling assessment by one of the great observers of American society, Alexis de Tocqueville. America thinks of itself as the most democratic of all nations. Shouldn't we have the most independence of mind and freedom of discussion? But Tocqueville argues **our respect for majority rule can discourage people from expressing unpopular ideas that run contrary to the majority's opinion.**

Tocqueville's point should concern leaders because if new ideas are not welcome, creative thinking becomes impossible. Let's consider four examples of how creative thinking is sometimes discouraged:

Advertising. Teens are constantly exposed to advertising. ***Branding – the process of associating certain visual, cultural, and even emotional images with a product – occurs when companies continually try to make their impression on us.***[52] None of the logos shown at right contains any words, yet you probably recognize most. Those symbols bring forth certain thoughts and feelings, demonstrating branding's effect on the mind. Out of habit, we reach for a particular cola. When we think of sneakers, a certain brand endorsed by a celebrity comes to mind. The success branding has had in shaping our behavior shows that advertising and popular culture sometimes stymie creative thinking.

Sedition Acts. The "Atlas of Independence," John Adams was present at the first Continental Congress. He helped Jefferson edit the Declaration of Independence, served as our first vice president, and later our second president. By any measure, John Adams was a leader in our nation's early history, a true champion of democratic values. And yet he signed into law the Alien and Sedition Acts, notorious laws "broad enough to make criminal virtually any criticism of the federal government."[54] ***Instead of embracing the open society, with its civic life marked by free inquiry and lively debate, Adams put his name to laws that mocked the First Amendment.*** Historians judge the Alien and Sedition Acts as a shameful chapter in U.S. history, illustrating that even America may erect barriers to free speech and creative thinking, despite our democratic aspirations.

Uniformity. Do you wear a uniform to school? The term "school uniform" brings to mind a certain outfit that is mandated by the school. But might the term mean more than that? Even if no particular outfit is officially required, teens often pressure one another to dress a certain way. Certain shoes are a must have, or a particular type of shirt is in style, and perhaps to be popular you have to wear your hair in a special way. This line of thought suggests that *so-called individualism or creativity may be conformity in disguise.* One social critic's perspective on tattoos illustrates the point:

> [The tattoo] is no longer a way to express individuality; it's a way to be part of the mob. People adopt socially acceptable transgressions — like tattoos — to show they are edgy, but inside they are still middle class. A cadre of fashion-forward types thought they were doing something to separate themselves from the vanilla middle classes but are now discovering that the signs etched into their skins are absolutely mainstream. They are at the beach looking across the acres of similar markings and learning there is nothing more conformist than displays of individuality, nothing more risk-free than rebellion, nothing more conservative than youth culture.[54]

The Military Tradition. In many ways, CAP resembles the military. We have a formal chain of command, dozens of regulations, strong traditions, and uniforms. As valuable as those features are, do they have an effect on creative thinking? According to one study, *the military lifestyle can brew hostility toward creative thinking.* Military officers generally score poorly on creativity tests, but score high in conformity tests. One observer noted that officers who are creative thinkers and non-conformists often do not advance in their careers because their embrace of creativity is mistaken for disloyalty or counter-conformity.[55] Therefore, leaders need to work extra hard to overcome a possible bias against creative thinking in military-style organizations.

> **"CAP and military leaders have to work extra hard to encourage creativity."**

TO CONSERVE, KEEP CHANGING

Leaders who value tradition and stability are apt to be resistant to change.

Indeed, one role of a leader is to protect what is positive and successful.

But does that philosophy require the leader to be skeptical about any kind of transformation?

G.K. Chesterton, the "Prince of Paradox," offers a surprising view:

"Conservativism is based upon the idea that if you leave things alone, you leave them as they are. But you do not.

If you leave a thing alone, you leave it to a torrent of changes. If you leave a white post alone, it will soon be a black post.

If you particularly want it to be white, you must be always painting it again . . . Briefly, if you want the old white post, you must have a new white post."[56]

- G.K. CHESTERTON

VICTORIES THROUGH CREATIVE THINKING

OBJECTIVE:

12. Give examples of outstanding creative thinking.

Returning again to our definition, creative thinking means seeing ideas in a new way, or connecting two concepts that seemed totally different.

Apollo 13. "We've got to find a way to make this fit into the hole for this, using nothing but that."[57] When a technical malfunction jeopardized the lives of Apollo 13's crew, the engineers and scientists supporting the astronauts knew that failure was not an option. But, despite their hundreds of checklists, contingency plans and back-up systems, there was no plan typed up on paper and neatly filed on a shelf to instruct NASA how to save the astronauts. The engineers would have to improvise, to think creatively, to make use of only the raw materials available to the astronauts. In the movie *Apollo 13*, the chief engineer is heroically defiant and proclaims, "This will be our finest hour."[58] And it was.

"Houston, we have a problem."
Thanks to quick thinking, creative engineers, Apollo 13 astronauts were able to build this contraption out of spare parts. Without it, they would have died in outer space.

D-Day. General Eisenhower planned WWII's D-Day invasion in great detail (see chapter 1). One half million troops and thousands of tons of equipment were to converge on the Normandy beaches with total precision. That was the plan at least. In reality, there was no order, only chaos and confusion. Soldiers who had been given orders to achieve certain objectives found themselves cut off from their units, in the wrong place at the wrong time, totally unable to execute Ike's well-orchestrated plan. What could our troops do? They improvised.[59] Sergeants and lieutenants hastily formed men from different units into teams, and set out to fight the Nazis as best as they could. In short, creative thinkers and leaders emerged just when we needed them most. Had our troops not responded with yankee ingenuity, the invasion would have failed, and perhaps Hitler would have won.

D-DAY, June 6, 1944, Omaha Beach.
Total chaos. No one landed where they planned. If not for creative, on-the-spot leadership by NCOs and low-ranking officers, the invasion would have failed.

Complexity & Simplicity.

"Simplify, simplify, simplify!" cried the poet Henry David Thoreau.[60] Must ideas be complex to be creative?

Consider "Rube Goldberg machines," the deliberately over-engineered, cartoonishly-complex inventions that through chain reactions accomplish a simple task. Indeed, it takes a creative mind and lively imagination to dream-up a great Rube Goldberg, and they're fun to watch.

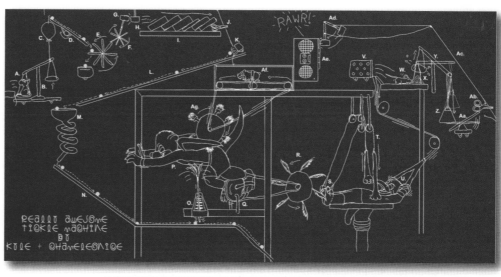

But the simple can be just as creative and ingenious. Shaker furniture, for example, is known for the beauty of its plainness. The Shakers were the polar opposite of Rube Goldberg. You will not find fancy decorations on anything made by a Shaker. They valued order, neatness, simplicity, and painted only in colors that would not attract too much attention.[61]

Creativity may be complex or simple.

Creativity: Complex and Simple.
A wacky drawing of a Rube Goldberg machine (top right) is an example of ultra-complex creative thought. In contrast, Shaker creativity took a much simpler form (right).

YOU MUST LIVE ON A BOAT!
More than 70% of the Earth's surface is covered by water.[62] Yet ironically, even the word "earth" means dirt, rocks, and land. Earth is more of a watery planet than an "earthy" planet. Might this image of the Earth (right), turned in such a way as to show hardly any land masses, remind us that sometimes we forget certain facts about our environment?

If this startling picture reminds us of something so basic about our home, imagine how many other obvious facts and preconceived notions have ingrained themselves into our minds. And might this image also suggest that what we see at first glance (in this case, a hemisphere that is almost totally water, with almost no land in sight) does not always give us a full and accurate understanding of reality? After all, more than 25% of the Earth is dry land, though you wouldn't know that from looking at this picture.

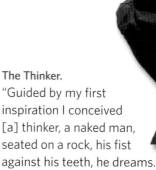

The Thinker.
"Guided by my first inspiration I conceived [a] thinker, a naked man, seated on a rock, his fist against his teeth, he dreams. The fertile thought slowly elaborates itself within his brain. He is no longer a dreamer, he is a creator."[63]

– AUGUSTE RODIN, French sculptor, 1840-1917

TOOLS FOR CREATIVE THINKING

OBJECTIVE:

13. Describe several practical methods of creative thinking.

How can leaders move their thinking out of the same old patterns? How can they look at problems in new ways? Outlined below are seven techniques to help you think creatively.

TOOLS TO GENERATE IDEAS

BRAINSTORMING

Purpose: To generate ideas through the quick, free-flow of thoughts

Procedure:[64]

1. Write the problem or topic where everyone can see it.

2. Include all ideas, do not edit the remarks.

3. Try to withhold judgment on the ideas. You want to generate ideas, not evaluate them.

4. Allow the team to have some quiet time just to think.

5. Try to involve everyone who has a stake in the problem. In brainstorming, the more people and perspectives you have, the better.

6. Be mindful that less assertive or lower-ranking cadets may hesitate to speak, for fear that their ideas are not good.

7. If the group is large, try breaking into small groups for the brainstorming. This helps generate even more ideas because each group is apt to attack the problem from a slightly different angle.

Example: How can we make encampment better?

More flying

Lower costs through sponsorships

Better staff training

Let cadet staff wear ascots, swords, berets, and leggings

Special tee shirts for each flight

Planning checklist for each staffer

Bunkmates not from same unit

MREs to eat in field

Do water survival at night, when you can't see anything

Scholarships for needy cadets

Arts & crafts supplies to make guidons

AFROTC cadets to visit for day

See if anyone knows how to bugle

Start blogging before arrival

Central point for everyone to upload photos

Basic, moderate, and tough levels for orienteering course

Add GPS navigation for hiking

Point system for honor flight

Do morning PT in uniform

More training for senior staff

Twitter updates to families back home

Make parents drop cadets off at gate and have cadets walk to command post

Have everyone bring a camelback for hydration

Make training time: lights out 12 midnight and reveille 4 am

Dance / party night before graduation

Notice that some of the ideas on this list probably aren't that great, but the recorder wrote them down anyway.

MINDMAPPING

Purpose: To generate new ideas in a creative way; to draw connections between different ideas

Mindmapping is a special way to do brainstorming. A mind map is a visual arrangement of ideas and their interconnections.[65] While a traditional outline represents linear thinking – one thing followed by another – mindmaps are radial, graphical, and non-linear.

Procedure:[66]

1. Position the main idea in the center. If a picture of that thing or idea is available, attach it.

2. Give yourself plenty of space (e.g.: large piece of paper or a whiteboard).

3. Look for relationships and draw lines between interconnected concepts.

4. Consider using different colors to illustrate the different themes that support the main idea.

5. Do not limit yourself to the use of words. Draw diagrams, attach pictures, graphs, or any objects that help you think about the subject.

Example: See below.

Proponents argue that mindmaps help connect left brain and right brain thinking.[67] As such, it can be a useful tool for generating new ideas, for taking notes during a lecture, planning a project, and most of all, for recording the results of a team's brainstorming session.

Surrealism, or The Wacky Door in the Sky. This school of art values surprise and juxtaposition. Surrealist paintings don't "make sense." Rather, the surrealists attempt to record thoughts that are uncontrolled by reason. What would surrealists think about mind maps?

Left Brain Thinking. Analysis, logic, science, math[68]

Right Brain Thinking. Emotions, intuition, art, creativity[68]

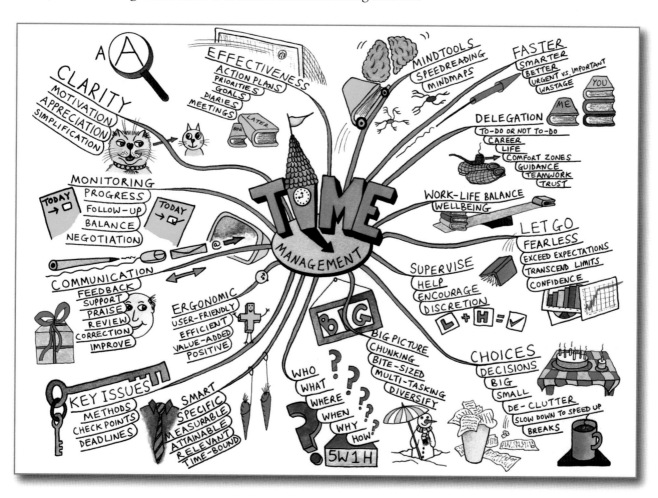

THE FIVE "WHYS"

Purpose: To discover new ideas and solutions to a problem by drilling down into a problem

Procedure: The team begins by stating the problem, then follows-up by continually asking "why"?[69]

Example: Our squadron is shrinking. We had twenty active cadets last year, and this year we have only twelve.

Why? Some cadets have sports?
  They practice Monday Wednesday and Friday and we meet on Monday
  We've always met on Monday
  Never thought of changing to Tuesday
  Take Away: Poll everyone to see if another day would be better

Why? Meetings are boring
  Too many lecture and people droning on
  No real resources available to do something else
  Take Away: Ask other squadrons what they've found successful and check CAP website for cool activities and training plans

Why? People just kind of fall away
  I don't know
  Never really asked them
  Not my job
  No one's told me to ask why other cadets quit
  Take Away: Assign someone to call cadets after they've missed 2 meetings, ask why they're not participating and invite them back

Why? Five cadets just graduated and are away at college
  Because when the 5 cadets left, they didn't get "replaced"
  We didn't recruit any "replacements"
  Not sure where to start to recruit new people
  No one's shown me how to recruit
  Take Away: Challenge every cadet to bring a buddy to CAP; give them recruiting materials; also, host an open house and put articles in the local paper to tell our town we're here

By continually asking "why?," the team can identify the root causes of a problem. Then, if the team can find ways to counter those root causes, there's a good chance they can solve the overall problem.

"I not only use all the brains that I have, but all the brains I can borrow."[70]
PRESIDENT WOODROW WILSON

Look closely because oddly enough, 21,000 soldiers got together one day to create this image of their commander-in-chief.

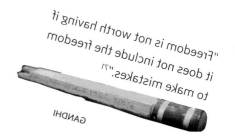

"Freedom is not worth having it it does not include the freedom to make mistakes."[71]
GANDHI

Creativity, the Air Force Way. Believing creative thinking is so important, the armed forces have invested millions in obstacle courses or leadership reaction courses, like the one these cadets are challenging.

REVERSAL

Purpose: To find a way to do something better; to improve a product or service

Procedure: Reversal is backwards brainstorming. Instead of trying to imagine ways to solve a problem, you imagine ways to create the problem.[72]

Example: Suppose you want to find better ways to welcome prospective cadets and help them earn their first promotion. To use the technique called reversal, you'd ask, "What could we do to make prospective cadets feel unwelcome, and how can we make it tough for them to become airmen?"

Don't introduce them to their squadron mates
Don't tell them what's exciting about CAP
Don't let them ask questions
Don't give them any training; make them learn the material on their own
Don't keep their parents informed of upcoming CAP activities

After exhausting their imagination, leaders work together to "re-reverse" their ideas. They'd make sure prospective cadets get introduced to their squadron mates, have someone tell them what's exciting about CAP, etc. The idea behind reversal is that to make something better it can help to ask what will make something worse.

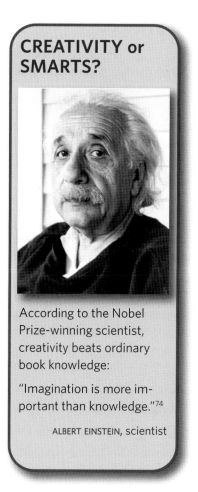

CREATIVITY or SMARTS?

According to the Nobel Prize-winning scientist, creativity beats ordinary book knowledge:

"Imagination is more important than knowledge."[74]

ALBERT EINSTEIN, scientist

HEADLINES OF THE FUTURE

Purpose: To analyze a problem and find the steps needed to achieve a goal.

Procedure: Imagine it's some time in the future and a journalist comes to report on your success in achieving a major goal. Ask yourself how the journalist's news story will read. Proceed to tell that story by actually writing it down as if you were a newspaper reporter. Be sure to include the technical details that made accomplishing your goal possible, and mention the smaller goals or milestones you had to achieve along your journey.[73]

Example: See right.

By using the "headlines of the future" technique, you are writing about your future picture (see chapter 2). ***Not only are you describing your dreams, you're analyzing what you must do to get there.***

DAILY NEWS

Former CAP Cadet Jane Curry Becomes Professional Pilot!

HOMETOWN USA -- Jane Curry has earned her Airline Transport Pilot rating from the FAA and is now flying as a co-pilot on a regional jet.

Curry qualified for the ATP rating by amassing over 1,000 hours as pilot-in-command and earning other ratings along the way including commerical, instrument, and multi-engine qualifications.

Curry's dream dates back to her early teens. She joined the Civil Air Patrol, became a cadet and studied aviation basics. At a CAP flight academy one summer, she flew her first solo.

Before graduating high school, Curry earned the Mitchell Award, which made her eligible for scholarships that helped her fund her college education and flight training.

FLOWCHARTS

Purpose: To identify the different parts of a system, and in so doing, to make it more efficient.

How can you get a handle on a big project that has lots of moving parts? A flowchart can help. *Flowcharts are simply visual representations of the major steps in a process.*[75] They help leaders find more efficient ways of accomplishing the mission. Moreover, they are often a good way to show everyone on the team (especially newcomers) what's going on.

Widget.
A fake, universal name for the thing an organization makes

Procedure:

1. Identify your widget. What is it exactly you're producing?

2. Describe the current process for making your widget. Where do you start? Then what happens? And then what? Include every step along the way (there are probably many more steps than you realize).

3. Step back from the flowchart and look for ways to make the route from "start" to "finish" easier.

4. Update your flowchart to show the changes you're making to your process.

Example: See right.

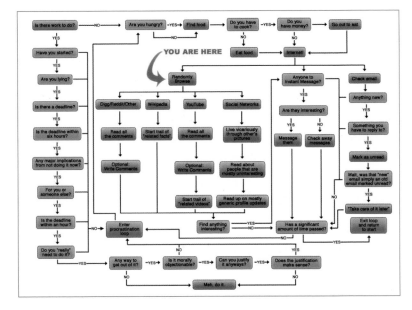

The best flowcharts are clear thinking made visible.[76] Special software is available to create flowcharts and share them online, but simply using a whiteboard, sticky notes, or tacking pieces of paper to the wall is sometimes an even better way to use flowcharts because it's so easy to update the chart, add little notes anywhere, and motivate the team as they track their progress.[77]

Cost / Benefit Analysis.
Another way to make decisions is to conduct a cost / benefit analysis.[79] In simple terms, you first consider what your ideas cost. These costs include money, time, effort, and the other opportunities you give up to pursue the goal. Then you consider the benefits, which could include making money, saving time, having more fun, learning something new, etc. In the end, you go with the idea only if its benefits outweigh its costs.

TOOLS TO MAKE A DECISION

MULTI-VOTING

Purpose: To find which idea has the greatest consensus when the team has several options to choose from.[78]

Procedure: Multi-voting is a democratic process, but instead of "one man, one vote," everyone gets to cast several votes. It's recommended that participants be given about half as many votes as there are options.

Example: All twenty cadets in the squadron are working together to choose goals for the coming year. They've made a list of 10 possible goals, and now want to identify the four goals that are most popular.

	1st 4 pts.	2nd 3 pts.	3rd 2 pts.	4th 1 pt.	Total	Rank
Rocketry	2	1	3	4	21	
Bivouac	5	2	2	3	33	3rd
Drill Team	3	3	2	4	29	4th
Orientation Flights	7	5	1	3	46	1st
Community Service Project	0	0	1	1	3	
"AEX Day" for Cub Scouts	0	0	1	1	3	
Radio Operator Training	1	1	1	0	9	
Orienteering Competition	1	7	4	1	34	2nd
Support Local Air Show	0	0	1	1	3	
Day Hiking (3 trips)	1	1	4	2	17	
Total Votes Counted	20	20	20	20		

Multi-voting avoids a win/lose situation for the team's members. It allows an item that was popular among most people, but not favored by all, to rise toward the top in popularity. In the example above, the orienteering competition was the favorite choice of only one cadet, but through multi-voting, the team realized that it was actually the second most popular overall.

WEIGHTED PROS & CONS

Purpose: To make a decision by analyzing the arguments for and against an idea, with a special emphasis on the relative strength of each pro and con.

Procedure:[80]

1. Make a chart that has 2 columns. Label one "pros" and the other "cons."

2. Under the "pros" column, list all the benefits you see your idea producing. Give each benefit you identified a point value. How many points you give the benefit is totally your decision.

3. Under the "cons" column, list all the drawbacks to your idea. This time give each a negative point value.

4. When you've completed your list, add up the points in each column. If the sum is positive, that means you're leaning toward going with the idea; if negative, you probably want to not adopt the idea.

Example: Should I get a part-time job so I can buy a car?

Pro	Con
It'd be awesome to have my own car; the general excitement of it +5	Cars cost a lot to buy -4
More freedom +3	Cost of insurance, gas, and maintenance -3
Gain valuable work experience +2	Having to work means less time for CAP -3
Be able to get to CAP activities without relying on anybody +1	Unnecessary; can use Mom's -2
TOTAL +11	Parents disapprove -2
	✓ TOTAL -14

MULTIVOTING:
Which activity is best?

With dozens of cool opportunities in CAP, how can a squadron decide which are its favorites? Multi-voting can help.

Flying

Rocketry

Orienteering

Bivouac

First Aid

Color Guard

Hiking

Radios

Air Show

What's the simplest way to make a decision? Add up the pros and cons. But because some pros and cons count for a lot and others matter only a little, *the weighted pros and cons method uses a point system to help you make decisions.*

GRADUAL VOTING

Purpose: Aid the team in making a sound and democratic decision by limiting the influence its ranking members have on the junior members.

Suppose you're somewhat against a proposal your team is considering. Before you have a chance to say why, the Spaatz cadet leading your team loudly says he is in favor of the idea. Might that intimidating cadet influence your vote? Might he or she deter others from speaking their minds? Gradual voting is a decision-making method that's useful when the team is comprised of members of different ranks or levels of experience.

> **"Gradual voting limits the influence senior people have over newcomers."**

Procedure: Begin by identifying the rank order of each member of the team (e.g.: lowest to highest ranking, youngest to oldest, or newest to most senior). In ascending order (lowest to highest), give each member a chance to share his or her perspectives on the following:

1.   State what the main issue is.

2.   Identify what factor affecting that issue is most important to you.

3.   Briefly summarize why you are in favor or against the proposal.

4.   Cast a provisional vote for or against the proposal, or if you have mixed views, explain how the proposal would have to be changed to win your vote.

5.   After giving everyone a turn, put the issue to an official vote by a simple showing of hands, to see if anyone has been persuaded to change their minds, based on what others have argued.

Example:

After reading written documents called "briefs," and listening to oral arguments, the nine justices of the Supreme Court of the United States meet in private to discuss the case and vote on the ruling. To help ensure each justice speaks his or her mind and votes honestly, the newest justice shares his or her views on the case first, then it is the second newest justice's turn, continuing up to the most senior justice. The one exception to this practice is that the Chief Justice, regardless of his or her seniority, speaks last.

Let the New Girl Go First. The Supreme Court of the United States uses gradual voting. The newest justice, #9, Elena Kagan, (as of this photo) goes first. Chief Justice John Roberts, #1, speaks and votes last.[81]

Through gradual voting, a leader can encourage every member on the team – especially the most junior – to truly speak their minds and vote accordingly. As a result, the team is apt to make better decisions.

TEACHING & TRAINING PEOPLE

"Education is not the filling of a bucket, but the lighting of a fire."[82]

WILLIAM BUTLER YEATS
Irish Poet

A fire begins with a tiny spark, but can quickly feed on itself and grow. No wonder that in art and literature, fire has long been a symbol for knowledge. Learn something today and that knowledge is sure to lead you toward learning something else tomorrow. What leaders do is something like lighting a fire, too. **Leaders try to influence other people, and in the process, make more leaders.** Recall that in chapter 3, we considered that one role a leader plays is that of the teacher. Leadership and education, therefore, go hand in hand. Both are like lighting a fire. Now that you are a cadet NCO, it's time to consider how you can become an effective teacher or trainer.

Light a Fire.
A fire begins with a tiny spark, but can feed on itself and grow. Several generations ago, we built warbirds. Today, Raptors. What will your generation build?

THE TRAINER'S GOALPOSTS

OBJECTIVE:

14. Explain the function of a learning objective.

What is it that a teacher or a trainer wants to accomplish? Is it something precise, or just a vague notion? In chapter two's discussion about goals, we learned that dreams can be vague, but goals have to be specific. **Likewise, effective teachers and trainers try to lead their people toward a precise goal called a learning objective.**

In simple terms, a learning objective describes what a student should know, feel, or be able to do at the end of the lesson.[83] The learning objective is the measure of success. If a student fulfills the learning objective, then it's mission accomplished. The student and instructor can move on to other challenges.

Good learning objectives are specific (saying exactly what is to be accomplished) and measurable (can be tested fairly).[84] To focus trainers and students on the specific and measurable, most learning objectives begin with action verbs, like these examples:

- Identify and describe CAP's four Core Values
- State the Air Force's definition of leadership
- Demonstrate how to perform the command, "To-the-Rear, MARCH"
- Find your position on the map by referencing nearby landmarks
- Show a commitment to Respect by properly rendering customs and courtesies

Note: Technically, each learning objective should begin with a phrase along these lines: "The objective of this lesson is for each student to..." That phrase is often taken for granted and so is sometimes not actually included with the stated objectives.

In other words, a learning objective tells the student and the instructor what the goal is. Moreover, the objective should be worded such that it's easy to tell if the student has fulfilled the objective or needs more help. As you can see, **without clear learning objectives, there's no way to tell if the lesson has been successful.**

Note: In high stakes tests, learning objectives are ultra-specific. In CAP, "Recite the Cadet Oath from memory" is a fine objective. But a hyper-precise objective might read: "Recite the Cadet Oath from memory, without help, and without omitting or reordering more than 3 words."

LEARNING STYLES

OBJECTIVES:
15. Describe the four modalities of learning.
16. Explain why leaders should present material in multiple ways.

How do you prefer to learn? By reading? By watching? By getting your hands dirty? *There are as many learning preferences as there are people. Everyone is different.* One way to understand this principle is to look at what educational theorists call the four modalities of learning. *The four modalities or learning channels describe the ways we process information into memory.*[85] They describe our learning styles or preferences.

Visual.[86] The visual cortex – the part of the brain controlling sight – is larger than all other sensory cortexes put together. No wonder then that many of us like to see what it is we're studying. Show me the aileron. Show me the cadet grade insignia. Show me what a flight in column formation looks like. *If you are a visual learner, you probably like to learn from watching videos, examining diagrams and looking at pictures.* You might like using flash cards to study and highlighting key passages in your textbook. Livedemonstrations make learning fun and productive for you.

Auditory.[87] Are you able to easily remember the words to your favorite songs? You may be *an auditory learner, someone who learns best by listening.* Story-telling and group discussions are effective ways to reach these learners. In grade school, the auditory learners probably liked hearing the multiplication tables being recited. Given a problem, an auditory learner's first instinct may be to talk about it. Or, given written instructions, the auditory learner is apt to ask that they be explained orally, thereby giving them a chance to listen. Perhaps its no surprise that auditory learners are likely to have developed an advanced vocabulary.

HOW DO PEOPLE REMEMBER BEST?

Research shows that on average, people retain:[87]

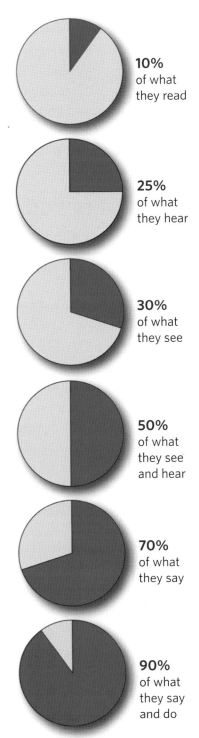

10% of what they read

25% of what they hear

30% of what they see

50% of what they see and hear

70% of what they say

90% of what they say and do

Tactile

Tactile.[88] The word *tactile* comes from the Latin word meaning "to touch." Therefore, **tactile learners will want to physically touch whatever it is they are studying.** Not content to simply see a handheld radio, for example, the tactile learner may want to try pressing its buttons, unscrew the antenna and screw it back on, open up the battery case and have a look around. In math class, you can spot the tactile learners because they're counting with their fingers. Labeling diagrams, drawing maps, or crossing items off a checklist are other hallmarks of tactile learning.

Kinesthetic

Kinesthetic.[89] This term comes from an ancient Greek word meaning "to move." Kinesthetic learning is closely related to tactile learning – both show a preference for learning by doing. But **with kinesthetic learning, the emphasis is on physically moving around and staying active, not simply touching things.** Games and role-playing are fun and productive ways to learn if you're a kinesthetic. If forced to sit through a lecture, the kinesthetic may want to use the computer to type their notes, that way at least their fingers stay busy. In cadet life, drill is an excellent example of kinesthetic learning. And in science class, the kinesthetic students will look forward to conducting experiments because they can get out of their seats and use their hands to learn.

Why is it important for leaders to know about the modalities? It's valuable to understand that everyone learns differently, so **whenever possible, instructors will want to present their material in a variety of ways, especially if the group is large.** And when giving instructions or mentoring someone, some basic knowledge about the modalities will help the leader communicate in a way that the follower/student finds easy to understand. Leaders who expect the team to adapt to their presentation style do not serve the team's needs.

> **"Leaders who understanding the learning modalities can communicate better."**

"Tell me, I forget. Show me, I see. Involve me, I remember."
CHINESE PROVERB

CHESS, WITHOUT ANY PIECES

An eight year-old chess prodigy is thinking deeply about a tough chess problem. Suddenly his trainer cries out, "Let me make it easier for you!," sweeping his arm across the board, knocking all the pieces to the ground.

The boy stares at the empty chessboard for a few moments. Then a light goes on in his brain: "Knight to A4!" Problem solved.

This scene from the film *Searching for Bobby Fischer* is a peculiar example of the four modalities of learning.[90] Peculiar in that the boy's pure genius allows him to learn without benefit of visual, auditory, tactile, or kinesthetic help.

How might we expect four chess players, each preferring a different modality, to train?

Visual: Would want to look at the piece and study the board; would do poorly in the scenario above.

Auditory: Would want to talk through the possible moves with the coach, or talk through the options to himself.

Tactile: Would struggle against a desire to unofficially move the pieces, which is not allowed.

Kinesthetic: Would prefer to get up, walk around the room, and perhaps look at the board from different angles.

TEACHING & TRAINING METHODS

OBJECTIVES:

17.   Describe different methods of teaching.
18.   Describe the pros and cons of those teaching methods.

Just as there are several modalities of learning, there are several teaching and training methods. No doubt, you've experienced each during your career as a student. But as you continue your transition from a follower to a leader, review the methods below with an eye toward understanding each from the instructor's point of view.

Lecture. *Arguably the most common teaching method, the lecture is an oral presentation of information, concepts, or principles that will lead students toward fulfillment of a learning objective.*[91] Lectures may be formal, with the instructor essentially reading from a carefully written manuscript, or informal with the instructor working from a rough outline but keeping the talk conversational in tone and occasionally welcoming students' questions.

Lecture

Instructors may overly rely on the lecture method because lectures are relatively simple to prepare. Moreover, they offer a quick way to impart large amounts of information, especially when introducing a new subject. It's an especially valuable method if the lecturer is an expert in the field who possesses knowledge that cannot be obtained elsewhere. But because the lecture is mostly a one-way form of communication, the lecture's main weakness is that it is difficult for the instructor and the students alike to tell if the students are actually learning the material. As such, lectures are a form of passive learning – the student is supposed to be like a sponge who absorbs the lesson, but in reality, students may simply sit back and let the instructor do all the work without listening critically. In a cadet environment, lectures are not very desirable because they are too "boring" and too much like school. Auditory learners are apt to enjoy lectures, but students who want to remain physically active as they learn will find them frustrating.

Guided Discussion

Guided Discussion. *The guided discussion is an instructor-controlled group process in which students share information and experiences to achieve a learning objective.*[92] While the instructor retains almost total control of the lesson during a lecture, in a guided discussion the students are supposed to do much of the talking. Therefore, the instructor's job is to facilitate or gently direct the conversation and challenge the students' remarks, leading them toward the learning objective. However, guided discussions can easily fail if the instructor cannot resist the urge to dominate the conversation, in which it becomes more of a lecture than a discussion. Because guided discussions usually do not require special equipment or involve elaborate activities, they are reasonably easy to prepare, though the instructor will still need to develop carefully worded questions in advance, along with ways to transition from one point to the next and summarize the most important teaching points. One assumption about guided discussions is that the students already possess some basic knowledge about the subject and are therefore capable of speaking intelligently about it. The guided discussion is a popular method because it provides the instructor with immediate feedback on the students' performance – you can tell who "gets it" based on how they are contributing to the discussion. Auditory learners will enjoy guided discussions, but students who are naturally quiet or prefer to be physically active while learning may find the method frustrating.

Demonstration – Performance. *The "demo-perf" is a process-driven approach to training that is used when students need to physically practice new skills.*[93] One assumption about demo-perfs is that most students learn best by doing. The classic example of proper use of the demo-perf is drill and ceremonies. The instructor demonstrates how to make an about face, then the students try it themselves. For the demo-perf to be successful, the instructor must be an expert in the subject. Not only must the instructor be able to demonstrate the task correctly, he or she must know the precise standards governing the task (exactly how far forward do you swing your arms on forward march?). Likewise, the instructor must be able to diagnose and correct students' problems in completing the task. Leading a demo-perf is easier than it looks – it actually requires outstanding communication skills, especially in a one-on-one setting. Kinesthetic and visual learners will love the demo-perf. One weakness of this method is that it works only for process-driven, task-based training – you can't teach someone to understand an abstract principle like "integrity first" via a demo-perf.

Experiential. *Experiential learning is an umbrella term covering games, role-playing, hands-on activities, service projects, problem-solving challenges, and more.*[94] With experiential learning, the main idea is to learn from doing, to learn from direct personal experience. This method often involves all modalities of learning, especially tactile and kinesthetic. Experiential learning has a reputation for fun and excitement and therefore most students are initially motivated to

Demo-Perf

SEATING ARRANGEMENTS

How an instructor sets-up the room can affect how the students relate with the instructor and with one another.

U-Shaped
Encourages students to interact with one another, not just the instructor; **ideal for guided discussions**

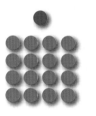

Traditional / Classroom
Interaction is between instructor and student; **ideal for lectures**

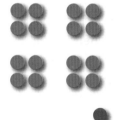

Independent Teams
Students interact almost exclusively within their own small group; **ideal for experientials and simulations**

participate, whereas students may be initially turned-off by a lecture, discussion, or other conventional method. One downside to experiential learning is that it almost always requires special materials and a great deal of planning and preparation. In cadet environments, experiential learning methods are especially important in ensuring CAP does not look and feel too much like school. Another potential downside to experiential learning is that it can be a victim of its own success. If learning is fun, the fun itself might become the goal, not a means to an end, which of course ought to be fulfillment of the learning objective.

Experiential

Simulation. A form of experiential learning deserving special attention is the simulation. *A simulation seeks to replicate the conditions of a job as realistically as possible.*[95] In CAP for example, we conduct search and rescue exercises where we simulate the actual tasks of aircrews, ground teams, and mission base staff in searching for a downed aircraft. Therefore, simulations are superb venues to practice existing skills and apply previously acquired knowledge. Generally the simulation is not a good method for teaching brand new concepts. The method is scalable, that is, it can be narrowly constructed and focused on teaching one very precise task (for example, pilots might use a computer-based flight simulator to practice recovering from spins), or the simulation may be wide-ranging in scope, encompassing all facets of a search and rescue mission, for example.

Simulation

To make the simulation as real-to-life as possible often requires a tremendous amount of visual aids, supporting materials, elaborate scripts, and the like. Having created the conditions necessary for running a simulation, the instructor's role becomes that of a mentor or evaluator. Simulations are great for tactile and kinesthetic learners. If the activity being studied is expensive or dangerous (e.g.: learning how to land an F-22 that has one engine out), simulations offer a cheap and safe alternative.

> "Simulations are superb venues to practice existing skills."

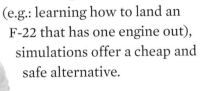

Experiential Learning
Leadership Reaction Courses are experiential learning. They require cadets to solve problems by applying knowledge acquired in the classroom.

TEACHING & TRAINING METHODS: PROS & CONS

Method	Pros	Cons
Lecture	★ Can present large amount of information ★ Taps personal knowledge and experience of instructor ★ Relatively easy to prepare	★ Communication flows mostly one-way ★ Boring, passive ★ Limited opportunity for feedback to check student learning
Guided Discussion	★ Students learn not only from instructor and from one another ★ Many opportunities for feedback and checking student learning ★ Students may relate topic to personal experiences	★ Facilitating a discussion and gently leading it requires skill and practice ★ Can easily fail if the group is too quiet or if instructor dominates discussion ★ Assumes students possess enough knowledge to speak intelligently about a subject; not appropriate for brand new topics
Demonstration Performance	★ Students learn skills by example and by actually practicing what they hope to learn ★ Great potential for individual attention ★ Great potential for immediate feedback, with students and instructor alike quickly seeing if objectives are being fulfilled	★ Not suited for in-depth academic study; designed for skills training ★ Labor-intensive, requiring lots of instructors / assistants, or a very small group size ★ Requires instructor to pay tremendous attention to detail and be skilled in diagnosing students' performance issues
Experiential	★ Active, exciting, fun way to learn ★ Generally involves all modalities of learning, thereby offering something for everyone ★ Offers great opportunities to synthesize learning across multiple topics or fields of study	★ Can become a victim of its own success, becoming an end in itself ★ Usually requires a great deal of instructor preparation, visual aids, and equipment ★ May convey only a small amount of actual academic content
Simulation	★ True to life, preparing students for the "real world" without experiencing its dangers ★ Active, exciting, fun way to learn ★ Great potential for feedback and learning from own successes and failures	★ Requires students to already possess basic knowledge; good for perfecting existing skills but not acquiring new ones ★ Usually requires a great deal of instructor preparation, visual aids, equipment, time, and other resources ★ Can often accomodate only a limited number of students at one time

RED FLAG: THE ULTIMATE SIMULATION

Nothing comes close to real combat. How then can Air Force warriors train effectively?

Red Flag is a combat simulation involving the full arsenal of US and allied aircraft.[96] Above the Nevada desert fly fighters, bombers, tankers, airlifters, helos, and nearly every kind of combat aircraft. For added realism, the "bad guys," called "Aggressors," are given unique paint schemes (left).

Operating on the ground in harsh conditions are the maintainers, medics, security forces, and other mission support personnel – all of whom must be self-supporting, just as if actually fighting in a distant land.

Military historians discovered that pilots' chances of survival dramatically increased after flying ten combat missions. Experience counts. Therefore, Air Force leaders created Red Flag to reproduce (as closely as possible) the life-or-death challenge of combat.

Red Flag offers the type of learning that cannot be acquired in a classroom.

EVALUATING LEARNING

OBJECTIVES:

19. Explain the purpose of evaluations, in the context of learning.
20. Give several examples of how cadets are evaluated.

Suppose you just completed your first challenge as a new instructor. You delivered a lecture, guided a discussion, conducted a demo-perf, or led some type of experiential activity. How do you know if you've been successful? It doesn't necessarily follow that what you present in class is what the student will learn.

In cadet training and educational settings, *evaluation is an attempt to check whether each student fulfilled the learning objectives.* Regardless how we try to measure learning, *an evaluation must be valid.*[97] That is, the test material must relate back to what the cadets studied. (Would you like to study land navigation but be tested on aircraft mechanics?)

Students may be evaluated formally or informally and in a number of ways, including: written tests, oral quizzes, participation in classroom discussions, and via direct observation.

> "Evaluations test whether students fulfilled the learning objectives."

After conducting an evaluation, there's still one step remaining: feedback. Just because you've completed some type of test or quiz doesn't mean you understand how well you did. *Through feedback, students see where they did well and where they fell short.* Feedback can be formal, such as when cadets see their written test results and correct their tests to 100%, or informal, such as when following a demo-perf, a cadet NCO says to an airman, "Not quite, watch my arms stay pinned as I perform the about face, see..."

CONCLUSION

In teaching and training people, you yourself become more knowledgeable and more expert in the subject matter. One great teacher wrote, "The least of the work of learning is done in the classroom."[98] The CAP Cadet Program develops leadership skills in cadets through textbooks like this one and classroom activities, but mostly through hands-on experiential learning. What's the best way to learn how to lead? By actually getting out there and using the cadet squadron as a leadership laboratory.

HOW CADETS MEASURE THEIR LEARNING

How many ways have you been evaluated as a cadet? Because evaluation is such an important part of teaching and training, there are several ways to measure learning, including:

Inspection

Written Exam

Drill Test

Discussion

Mile Run

Flying Solo

THE CAP CADET PROGRAM'S LEARNING MODEL

The CAP Cadet Program is a good example of "student-centered education." That is, the focus is on the cadets' needs, interests, and abilities. This table shows how the Cadet Program applies the student-centered learning.[99]

Theory	Practice
1. Learning is most meaningful when topics are relevant to the students' lives, needs, and interests . . .	1. Cadets join CAP because they want to fly, learn about the military, or for similar reasons.
2. . . . and when the students themselves are actively engaged in creating, understanding, and connecting to knowledge.	2. Cadets learn by doing. They learn about aviation by flying in CAP aircraft. They learn to lead by serving on a cadet staff.
3. Students will have a higher motivation to learn when they feel they have a real stake in their own learning.	3. Cadets are eager to advance so they can earn promotions, ribbons, and awards, and qualify for prestigious staff positions.
4. Instead of the teacher being the sole, infallible source of information, then, the teacher shares control of the classroom and . . .	4. Ranking cadets instruct, train, an mentor junior cadets, under the guidance of a senior.
5. . . . students are allowed to explore, experiment, and discover on their own.	5. Cadets have opportunities to participate in special activities, but are not required to do so. There's something for everyone in CAP.
6. Essentially, learners are treated as co-creators in the learning process, as individuals with ideas and issues that deserve attention and consideration.	6. The cadet staff has a say in the goals the squadron sets. They help plan and implement cadet activities, under senior guidance.

CHECKLIST: HOW TO LEAD A DEMO-PERF

When teaching someone how to perform a task – how to drill, how to use a compass, how to preflight an airplane – the demonstration / performance method can be a great way to train. Here's how it works, as applied to drill:

1. State the movement and explain its purpose.

2. Perfectly demonstrate how the movement is performed at a normal cadence, twice.

3. Break the movement into segments. Show the starting position and the finishing position. Identify any special rules or standards. Slowly demonstrate the movement one step at a time, by the numbers. Allow the cadets to ask questions.

4. Have cadets try executing the movement on their own, and then as a group, by the numbers. Watch them closely and give them feedback. Ensure everyone understands how to perform the movement properly.

FINAL ANALYSIS

Leadership requires brainpower. Deep, serious, ever-growing brainpower. Aspiring leaders can develop their brainpower by studying principles of critical thinking and by learning how to be more creative. Moreover, brainpower is especially important in the leader's role as an instructor.

Our study of leadership keeps returning to the principle that leadership is an intellectual activity. Great leaders are great thinkers. Therefore, any cadet who means to lead must develop his or her brainpower.

DRILL & CEREMONIES TRAINING REQUIREMENTS

As part of your study of this chapter, you will be tested on your ability to lead an element in drill and ceremonies. Ask an experienced cadet to help you develop your command voice and practice calling commands on the correct foot. For details, see the *USAF Drill and Ceremonies Manual* available at capmembers.com/drill.

From the Air Force Drill & Ceremonies Manual, Chapter 4

Command the flight to fall in.

Command the flight to dress right and check its alignment.

Command the flight to open and close ranks and check its alignment.

Command the flight to perform facings and other in-place movements.

Command the flight to perform flanks, columns, and march to-the-rear.

Command the flight to perform right (left) steps.

Command the flight to perform close and extend, at the halt and on the march.

Command the flight to change step and count cadence.

Command the flight to form a single file or multiple files.

ENDNOTES

1.   John C. Maxwell, *How Successful People Think*, (New York: Hachette, 2009), x.

2.   Linda Elder, "Defining Critical Thinking," (The Critical Thinking Community, 2007), http://www.criticalthinking.org, retrieved Dec 2009.

3.   Plato, Apology, 38a.

4.   Elder.

5.   Jacques Louis David, "Death of Socrates."

6.   Michael Scriven & Richard Paul, "Critical Thinking as Defined by the National Council for Excellence in Critical Thinking," (The Critical Thinking Community, 1987), http://www.criticalthinking.org, retrieved Dec 2009.

7.   "Word of the Year 2006," http://www.merriam-webster.com/info/06words.htm, retrieved Nov 2009.

8.   Nathan Rabin, "Interview: Stephen Colbert," (*The Onion*, Jan 06), http://www.theonion.com, retrieved Dec 2009.

9.   Elder.

10.   Edward Tufte & Jimmy Guterman, "How Facts Change Everything," *Sloan Management Review*, Summer 2009, 35-38.

11.   Maxwell, 5.

12.   Marcus Aurelius, *Meditations*, II.9.

13.   Alvin Toffler, attributed.

14.   Voltaire, attributed by Creating Minds.org, http://creatingminds.org/quotes/thinking.htm, retrieved Nov 2009.

15.   Harry Overstreet, *The Mature Mind*, (New York: Norton, 1949).

16.   Maxwell, 14.

17.   Glenn Wilson, King's College London University, http://www.prweb.com/releases/2005/04/prweb232729.htm, retrieved Dec 2009.

18.   Maxwell, 19.

19.   Maxwell, 38-39.

20.   Ibid, 40.

21.   Michael Waldman, *My Fellow Americans*, (New York: Sourcebooks, 2003) 137.

22.   Stephen Lorenz, "Lorenz on Leadership," *Air & Space Power Journal*, Summer 2005.

23.   Maxwell, 41.

24.   Ibid, 93.

25.   Ibid, 94.

26.   Catherine Drinker Bowen, Miracle at Philadelphia, (Boston: Back Bay, 1986 ed.).

27.   Maxwell, 97.

28.   Ibid, 97.

29.   Attributed by Maxwell, ix.

30.   Ronald Reagan, "First Inaugural Address," 1981.

31.   Michael Labossiere, "Fallacy Tutorial Pro," at The Nizkor Project, http://www.nizkor.org/features/fallacies/, retrieved Dec 2009.

32.   Ibid.

33.   Thomas J. Watson, "Comments on Think," IBM Archives, http://www-03.ibm.com/ibm/history/multimedia/fulldescriptions/think.html, retrieved Dec 2009.

34.   Jeremiah Denton, *When Hell Was in Session*, (New York: Reader's Digest Press, 1976).

35.   Bobzien, Susanne, "Ancient Logic", *The Stanford Encyclopedia of Philosophy*, http://plato.stanford.edu/archives/fall2008/entries/logic-ancient/, retrieved Nov 2009.

36.   Attributed

37.   Colin Raye, "Not That Different," *I Think About You*, [CD] Sony: 1995.

38.   Jim Arvo, "On Intellectual Honesty," (UC Irvine), http://www.ics.uci.edu/~arvo/honesty.html, retrieved Dec 2009.

39.   Ibid.

40.   Fyodor Dostoevsky, *The Brothers Karamazov*, trans. Richard Pevear & Larissa Volokhonsky, (New York: FSG, 2002), II.2.

41.   Unknown.

42.   Ralph Waldo Emerson, "Self Reliance," 1841.

43.   George Bernard Shaw, *Man and Superman*, 1903.

44.   Dr. Seuss, *Oh the Thinks You Can Think!* (New York: Harper Collins, 2000), 1.

45.   Talking Heads, "Memories Can't Wait," *Fear of Music*, (New York: Sire, 1979).

46.   John A. Warden III, *Winning in Fast Time*, (Montgomery, AL: Venturist, 2002), 72.

47.   Ibid, 87.

48.   Tucker Automobile Club of America, "Tucker History," http://www.tuckerclub.org, retrieved Jan 2010.

49.   Ralph Nader, *Unsafe at Any Speed*, (New York: Grossman, 1965).

50.   Ibid.

51.   Alexis de Tocqueville, *Democracy in America*, trans. Henry Reeve, I.15, (1835).

52   "Branding," http://www.businessdictionary.com/definition/branding.html, retrieved Nov 2009.

53.   Richard Labunski, *Libel & the First Amendment...*, (New York: Transaction, 1987), 35.

54.   David Brooks, "Nonconformity is Skin Deep," *New York Times*, Aug 27, 2006.

55.   Lt Cmdr Anthony Kendall, "The Creative Leader," in Concepts for Air Force Leadership, (U.S. Air Force: Maxwell AFB, AL, 196), 309.

56.   Attributed by Rich Lowry, *National Review*, March 4, 2009.

57.   *Apollo 13*, directed by Ron Howard, (Hollywood: Universal, 1995).

58.   Ibid.

59.   Stephen E. Ambrose, *D-Day*, (New York: Simon and Schuster, 1995).

60.   Henry David Thoreau, *Walden*, 1854.

61.   "About the Village," http://www.shakers.org, retrieved Nov 2009.

62.   "Water Science for Schools, " US Geological Survey, http://ga.water.usgs.gov/edu/earthwherewater.html, retrieved Dec 2009.

63.   Artcyclopedia, http://www.artcyclopedia.com/feature-2001-08.html, retrieved Nov 2009.

64.   "Brainstorming," MindTools, http://www.mindtools.com, retrieved Dec 2009.

65.   "Mind Maps," MindTools, http://www.mindtools.com, retrieved Dec 2009.

66.   Ibid.

67.   Ibid.

68.   "Right Brain vs. Left Brain," Funderstanding, http://www.funderstanding.com/content/right-brain-vs-left-brain, retrieved Dec 2009.

69.   Air Force Quality Institute, *Process Improvement Guide*, (Maxwell AFB, AL: Air University, 1994), 8.

70.   Woodrow Wilson, "Remarks to the National Press Club," 20 March 1914 in Arthur Link, ed., *The Papers of Woodrow Wilson*. (Princeton, NJ: Princeton University Press, 1979) 29:363.

71.   Gandhi, attributed.

72.   "Reversal," MindTools, http://www.mindtools.com, retrieved Dec 2009.

73.   Warden, 63-69.

74.   George Sylvester Viereck in *The Saturday Evening Post* Vol. 202 (26 October 1929), p. 117

75.   "Flow Charts," MindTools, http://www.mindtools.com, retrieved Dec 2009.

76.   Edward Tufte, quoted in *Hartford Advocate*, October 1999.

77.   Ibid.

78.   Air Force Quality Institute, 16-17.

79.   "PMI," MindTools, http://www.mindtools.com, retrieved Dec 2009.

80.   Ibid.

81..   Alan Green, *A Guide to the Supreme Court...*, (Washington: Benton Foundation, 1987), 36.

82.   Attributed to Yeats.

83.   U.S. Air Force, AFMAN 36-2236, *Guidebook for Air Force Instructors*, ch. 3.

84.   Ibid.

85.   College of Charleston Center for Student Learning, "Learning Styles," http://spinner.cofc.edu/studentlearningcenter/?referrer=webcluster&, retrieved Dec 2009.

86.   Ibid.

87.   Ibid.

88.   Ibid.

89.   Ibid.

90.   *Searching for Bobby Fischer*, directed by Steven Zaillian, (Hollywood: Mirage, 1993).

91.   U.S. Air Force, AFMAN 36-2236, *Guidebook for Air Force Instructors*, ch. 13.

92.   Ibid, ch. 14.

93.   Ibid, ch. 17.

94.   Ibid, ch. 12.

95.   Ibid, ch. 12.

96.   Nellis Air Force Base, "Red Flag," http://www.nellis.af.mil/library/flyingoperations.asp, retrieved Dec 2009.

97.   U.S. Air Force, AFMAN 36-2236, *Guidebook for Air Force Instructors*, ch. 22.

98.   Attributed to Fr. Louis Merton OCSO (Thomas Merton).

99.   Wisconsin Center for Educational Research, "Student-Centered Learning," http://www.wcer.wisc.edu/, retrieved Dec 2009.

CHAPTER 6
THE HUMAN ELEMENT

HOW CAN WE UNDERSTAND LEADERSHIP WHEN WE CAN HARDLY UNDERSTAND PEOPLE? The great variable in the equation of leadership is the human element. Humankind has set foot upon the Moon, explored Mars, probed the solar system and beyond. And yet, nothing is so puzzling as what is immediately before us: the human mind. Shakespeare summed it up when he wrote, "What a piece of work is man!" Everyone is alike, and yet everyone is different. People are the problem, and people are the solution. People cause suffering. They hate. To get some to work, you must kick them out of bed. But people are Earth's most warmhearted and hopeful animals. They built civilization, uncovered mysteries of science, and brought art and music to a cold universe. People are a paradox. Humankind, wrote one poet, is the "glory, jest, and riddle of the world!"

PERSONALITY

OBJECTIVE:
1.   Define the term, "personality."

Everyone is the same, and yet everyone is unique. This is the mystery of personality. No doubt you know someone who is said to have an outgoing personality, or another who is known for having a serious personality. But what is "personality"? In simple terms, *personality is the sum of the thoughts, feelings, and behaviors that make someone unique.*[1] Some features of personality are visible to all – it's easy to find the clown in the group – while other features of personality lay under the surface, hidden from the outside world and perhaps even hidden from the individual himself.

What factors shape personality? How can we better describe personality? This section considers those questions and more. Leaders try to understand personality so that they might better understand people.

NATURE VS. NURTURE

OBJECTIVES:
2.   Describe ways that nature influences personality.
3.   Describe ways that nurture influences personality.
4.   Explain why the nature vs. nurture debate is relevant to leaders.

Are you the way you are because you were born that way? Have your genes determined the type of person you have become? Is biology destiny? Or, have you been formed by your personal experiences? Have your parents, friends, school life, and the like molded you into the person you are? Welcome to a debate that is over 400 years old. In short, *the question of "nature vs. nurture" asks whether it is inborn qualities or personal experiences that shape who we are.* The classic question, "Are leaders born or are they made?" is closely related to the nature vs. nurture debate.

CHAPTER GOALS

1.   Develop an understanding of what makes individuals unique and complex.

2.   Appreciate how interpersonal relations affect the job of leading.

3.   Defend the idea that diversity is a strength.

CHAPTER OUTLINE
In this chapter you will learn about:

Personality
   Nature vs. Nurture
   Birth Order Theory
   Charisma
   Johari Window
   Myers-Briggs Type Indicator

Motivation & Behavior
   Maslow's Hierarchy of Needs
   Hawthorne Studies
   Classical Conditioning
   Milgram Experiment

Conflict
   Defense Mechanisms
   The Inevitability of Conflict
   The Leader's Role in Managing Conflict

Leading in a Diverse Society
   Diversity in the Military & CAP
   America's Increasing Diversity
   Prejudice, Hatred, & the Leader
   Five Ways to Fight Hate

Drill & Ceremonies

ARGUMENTS FOR NATURE

Consider a pair of identical twins. Biologists tell us that identical twins possess identical genes. Researchers have found that if identical twins are raised apart in different families, they will nevertheless grow up to be highly similar.[2] It seems that nature has programmed them a certain way. Although each twin was nurtured by different parents, nurtured at different schools, and nurtured by different friends and family, nature still found a way for the twins to grow up to be very much alike. Nature affected their personality, intelligence, interests, individual quirks and more.

> **"Are you the way you are because you were born that way? Or have you been formed by personal experience?"**

Physical traits such as eye color, hair color, height, weight, and the like are controlled by nature. Geneticists can tell a couple the likelihood of their children having blue eyes or brown. Down syndrome, cystic fibrosis, Huntington's disease, and hundreds of other medical problems are the result of genetic disorders. In short, our parents' DNA determines a great deal about who we are.[3] No matter how carefully parents nurture a child, the laws of genetics will have their way.

Twins
Identical twins have identical genes. Indeed, they're so alike that researchers have found that if raised separately, each twin is apt to grow up to become much like the other. It seems that nature is in control.

ARGUMENTS FOR NURTURE

Consider two puppies. Put one through obedience training and do nothing to train the other. It will be no surprise which dog learns to sit, stay, and lay down, and which is utterly unable to perform at the same level. Nurturing has an effect. The same principle holds true for violence. Researchers have discovered that children who grow up around violence are apt to become violent themselves.[4]

The argument for "nurture" is best expressed by the concept of the blank slate, or *tabula rasa*, as it is called in Latin. ***The blank slate principle states that every newborn baby is born as if their mind were a blank slate onto which they write thoughts and experiences.***[5] This argument asserts that we take-in information using our senses and are formed by our life's events. "Man has no nature," announced one writer, "what he has is history."[6]

One researcher, John B. Watson, was so steadfast in his belief that nurture overpowers nature, he famously proclaimed:

> Give me a dozen healthy infants, well-formed, and my own specified world to bring them up in and I'll guarantee to take any one at random and train him to become any type of specialist I might select... regardless of his talents, penchants, tendencies, abilities, vocations and race of his ancestors.[7]

Good Girl!
What makes dogs able to sit, stay, and fetch on command? Training. Nurturing can produce powerful effects.

Tabula rasa.
Latin for "blank slate."

NATURE VS. NURTURE TODAY

Today, most scientists reject the nineteenth-century doctrine that biology is destiny and the twentieth-century doctrine that the mind is a blank slate.[8] ***We are affected by our genes. We are affected by our environment.*** "The brain," explained one scientist, "is capable of a full range of behaviors and predisposed to none."[9] Nature and nurture are not mutually exclusive. Rather, nature and nurture affect one another.

IMPLICATION FOR LEADERS

What does the nature vs. nurture debate mean for leaders? Regardless of whether nature dominates nurture or vice versa, we know that ***leaders cannot change human nature.*** But we also know that a person's environment has an effect on how they develop, and ***a leader can have an effect on that environment.*** The key ingredient in the leader/follower environment is the leader's own behavior. Once again the simple wisdom, "lead by example," is shown to be leadership's first commandment.

Nature vs. nurture teaches us that everybody is alike, and yet everybody is different. The wise saying, "know your people" comes round again. Cadet NCOs leading small teams must get to know their people as individuals. Only then can they discover what it will take to nurture, support, and lead those individuals.

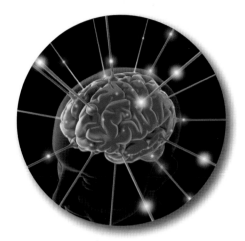

The Human Brain
According to one scientist, "the brain is capable of a full range of behaviors and predisposed to none."[8]

NATURE VS. NURTURE AT THE MOVIES

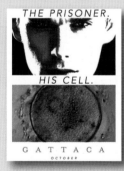

The film *Gattaca* imagines a future where your genes are your destiny. Seconds after birth, each baby is genetically tested. It is at this moment that the newborn's station in life is determined. The babies with the best genes are selected to become astronauts. Those with inferior genes are slated to become janitors. No matter how hard you work, how much you study, how many push-ups you can endure, in Gattaca society refuses to believe you can succeed if you have bad genes. Gattaca is a world where nature appears totally in control.[10]

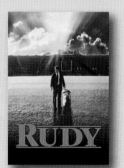

Rudy Ruettiger stands 5'7" tall, tiny for a football player. Possessing average intelligence, he was not smart enough to go from high school straight into a prestigious college. The film *Rudy* tells the story of how through sheer willpower and nurturing, Ruettiger achieved his dream of playing football for the Irish and graduating with honors from Notre Dame.[11] Rudy's story rebuts the claim that nature trumps nurture. Today in real life, Ruettiger is a famous leadership coach. "Be the person you want to be," he advises. "Make the decision to take action and move closer to your dream."[12]

BIRTH ORDER THEORY

OBJECTIVE:

5.   Describe birth order theory.

Birth order theory contends that a person's rank within their family can have an effect on their personality and intelligence. It is the idea that all first born children, for example, will hold certain experiences in common and that those experiences will mold their personality in a predictable way. Where a child places in the birth order – first born, middle, youngest, or only child – can have an effect on how he or she sees themselves.[13]

ARGUMENTS FOR BIRTH ORDER THEORY

Researchers have uncovered some data to support the birth order theory. For example, one survey found that 43% of company presidents (not 33% as would be expected) were first born children, suggesting that being a leader of a little brother or sister may equip individuals to be leaders in adult life.[14] Other researchers have found that first born children are over-represented in Congress.[15] In contrast, a disproportionate number of last born children have been arrested as protesters, suggesting that the youngest members of a family are apt to be the most rebellious.[16]

> **"Will all first born children become successful? Will the last born rebel?"**

Many people find birth order theory persuasive because it agrees with their common sense. The "tutor effect" is a good illustration.[17] A first born child, for example, will have ample opportunity to develop leadership skills by acting as a tutor or boss to the younger siblings. Big brothers and sisters teach their little brothers and sisters how to tie shoes, throw a baseball, apply make-up, or add and subtract. In Norway, scientists attribute the tutor effect to their discovery that first born children have an IQ 2.3 points higher, on average, than children born second or last in their families.[18]

ARGUMENTS AGAINST BIRTH ORDER THEORY

But birth order theory has its opponents, too. Some scientists equate it with astrology, palm reading, pop psychology, or the like.[19] They accuse it of the *post hoc* fallacy (see chapter 5). While birth order may have some effect on how we see ourselves and how personality develops, opponents argue that other factors are more important. The timing of economic shocks to the family, the timing of the family moving to a new city, or the timing of any number of major life events offers a better explanation of how personality develops than birth order does.[20]

THE PRESIDENTS & BIRTH ORDER

Eight of the past ten presidents can be considered first born children. Advocates of birth order theory claim first borns have an edge in developing leadership skills.

44. Barack Obama
Only child

43. George W Bush
Oldest child

42. Bill Clinton
Oldest child*

41. George H Bush
Second child

40. Ronald Reagan
Last born

39. Jimmy Carter
Oldest child

38. Gerald Ford
Oldest child*

37. Richard Nixon
Oldest child✢

36. Lyndon Johnson
Oldest child

35. John Kennedy
Oldest Child✢

✱ *Only child, but later became the oldest child in a mixed family*
✢ *Born second, but older brother died at a young age*

RELEVANCE TO LEADERS

Why is birth order theory relevant to leaders, if at all? What are we to make of birth order theory when some scientists argue for it and others against it? ***Birth order theory is an easy way to begin thinking about how the environment a person grows up in can shape their personality.*** A familiarity with birth order theory may help a leader better understand why someone displays the personality traits that they do. It is yet another item in the leader's toolkit to give them a larger perspective about people.

TYPICAL CHARACTERISTICS OF INDIVIDUALS ACCORDING TO BIRTH ORDER THEORY[21]

Possible Leadership Approaches*

Only Child

Spoiled

Used to getting all the attention

Self-centered and used to getting their own way

Over-relies on authority figures (e.g.: parents)

Feels stupid in early childhood because adults are always more capable

Not good at cooperating because they do not need to as much

Appeal to their sense of having special qualities and deserving special attention. "Cadet Curry, your color guard experience can really help us here..." Challenge them to see how their personal goals match the team's goals.

First Child

Displays only child characteristics, at least for a while

Focuses on having control or authority over the younger children

Enjoys being right and having more knowledge and experience than the others

May feel jealous and neglected when second child is born

Strives to please

Strives to achieve and feels pressure to succeed

Appeal to their desire to please and achieve. Challenge them to excel. "Cadet Curry, I just know you're capable of passing the challenging Wright Brothers exam, will you study hard next week so we can promote you to C/SSgt?

Middle Child

Frustrated due to not having the special qualities of being oldest or youngest

Feels left out, unloved, and forgotten

Feels life is unfair and that they receive only hand-me-downs

Rebels against perceived injustices

Seeks an identity outside the family

Learns the necessity of compromise

Struggles in always being compared to the oldest child

Emphasize their sense of belonging to the team and the special honor of being a cadet. Challenge them to channel their rebellious attitude. Ask them to be innovators who discover new ways for the team to succeed.

Youngest Child

Feels inferior to the bigger, smarter, stronger, older siblings

Remains "the baby," even into adulthood

Expects to be cared for and expects others to take responsibility for them

Behaves like the only child, at least for a time

Frustrated by a feeling they are not respected or taken seriously

Used to being the center of attention

Feels left out from the family for missing experiences from before their time

Appeal to their sense of having special qualities and deserving special attention. "Cadet Curry, your color guard experience can really help us here..." Take them seriously and challenge them to make their mark by doing something extraordinary for the team.

★ *The goal of this section on birth order is to help you understand that a person's rank within their family can have an effect on their personality. In most cases, especially in the adult workplace, leaders will not know the birth order of their followers.*

CHARISMA

OBJECTIVES:
6.   Define the term, "charisma."
7.   Explain why charisma can help a leader succeed.
8.   Explain pitfalls leaders can face by relying too much on charisma.

"Charisma is the sparkle in people that money can't buy," according to one author. "It's an invisible energy with visible effects."[22] Someone who has great charisma might be described as having a magnetic personality, a unique flair, a special quality that is hard to describe and even harder to imitate. Charismatic leaders have a profound emotional effect on their followers. In short, charisma is that special aspect of personality that makes someone truly unique.

WINNING HEARTS AND MINDS

What does charisma mean for the leader? Charismatic leaders find it easy to recruit new followers. In politics, candidates for office go to great lengths to appear likable. By putting their charisma on display, they hope to win voters. Moreover, because charismatic leaders have by definition a "profound emotional effect," their followers will work longer and harder for the cause.[23] Having a strong charisma can make a leader appear heroic and larger-than-life, and therefore incredibly effective.[24] Name some beloved leaders – Martin Luther King, Ronald Reagan, Oprah Winfrey – and you'll be naming individuals whose success is due partly to their extraordinary charisma.

A PERILOUS SHORTCUT

Having good charisma, a sunny disposition, and being likable certainly makes leadership easier for the leader. But charismatic leadership is not necessary for an organization to succeed.[25] *Moreover, strong charisma can be counter-productive as it surrounds the leader with followers who are only too willing to flatter the leader and sweep problems under the rug.*[26] "Charisma becomes the undoing of leaders," warns one expert. "It makes them inflexible, convinced of their own infallibility, unable to change."[27] By being zealously loyal to the leader, both the followers and the leader alike can forget their duty to be loyal to the organization's goals or core values. And when viewed from the perspective of a cadet searching for ways to develop leadership skills, charisma is a blind alley. You cannot make yourself more charismatic and stay true to yourself. Regardless, what one person finds attractive about a leader's charisma might turn-off someone else.

THE RESULT OF LEADERSHIP

Is charisma the X-factor? Is it a magic potion that turns mere mortals into leaders? No. One expert said it best: "Charisma is the result of effective leadership, not the other way around."[28]

BILL CLINTON: THE PROS & CONS OF CHARISMATIC LEADERSHIP

Nobody works a crowd as well as former president Bill Clinton, one of America's most charismatic leaders.

One journalist observed that Clinton had an uncanny ability to charm people. "When he shook a person's hand, he leaned slightly forward, looked that person in the eye, and made that individual feel like he or she was the only person in the room."[29] Clinton's charisma influenced people, built support for his programs, and won elections.

But there was a dark side to that charisma. In a major scandal, he was accused of perjury. Clinton assured his cabinet that he was innocent, so they supported him.[30]

When the truth finally came out and it was clear Clinton had lied under oath, his cabinet realized they had been betrayed.

Would his closest advisors have believed him had not his charisma been so powerful? What does the Clinton presidency teach leaders about the capabilities and dangers of charismatic leadership?

CHARISMATIC LEADERSHIP: TWO VIEWS

The Horror of Jonestown

What would it take to persuade you to sell all your belongings, leave your family, and move to a tiny camp of primitive huts in the jungle of South America? In the 1970s, one leader of extraordinary charisma, Jim Jones, was able to convince hundreds of people to do just that.

At first, Jonestown, as it was called, was a great success. A mesmerizing speaker, Jones used his incredible charisma to establish what seemed like the most peaceful, loving community on earth. People who had been miserable all their lives found happiness with Jones as their leader.

And then, tragically, one day Jones turned his charisma toward evil ends. Acting on his instructions, hundreds of his followers calmly and senselessly filled their cups with poison-laced kool-aid, drank it, and died.[31]

At its worst, charisma, according to one expert, is based in worshipful emotions of devotion, awe, reverence, and blind faith.[32] A charismatic leader may find it easy to prey on people's weaknesses. The horrors of Jonestown illustrate that charismatic leadership can be deadly.

Forrest Gump and Charisma

There is nothing Forrest Gump would not do for his leader and friend, Lieutenant Dan.

At first, we have a hard time figuring out this relationship. Lieutenant Dan is just plain mean toward Forrest. He makes Forrest the butt of his jokes. No matter. Forrest responds by being even more kindhearted, and yet Lieutenant Dan makes fun of Forrest some more.

Why is this? As the film progresses, we see that Forrest admires Lieutenant Dan because of, not in spite of, his weird personality. In Forrest's eyes, Lieutenant Dan has charisma.

Murderous Charisma
Over 900 people senselessly ended their lives in an act of blind devotion to their leader, Jim Jones, in 1978.

JOHARI WINDOW

OBJECTIVES:
9.   Describe the four arenas of the Johari window.
10.   Explain how feedback and self-disclosure help leaders.

Do you know someone who has delusions of grandeur? Perhaps a cadet believes she is destined to attend the Air Force Academy, and yet her fellow cadets recognize she simply doesn't have the self-discipline to succeed or be happy at a service academy. At some point, reality will catch up with the cadet. Her "blind spot" is hindering her ability to reach her goals.

The Johari Window is a tool for exploring our self-perception.
What do we know and not know about ourselves? What do others know and not know about us? The Johari Window offers a way for exploring those questions. Its unusual name is a combination of the first names of the researchers who developed its framework, Joseph Luft and Harrington Ingham (Joe + Harry).[34]

FOUR ARENAS

The "window" is a model that is divided into four "arenas."[35]

	Known to self	Not Known to self
Known to Others	Public	Blind
Not Known to Others	Private	Unknown

Diagram of the Johari Window

The public arena consists of those features of your personality that you know about yourself and that others know, too. Those features become public thanks to good self-perception coupled with good communication.

The blind arena consists of those personality features that are unknown to you but are known to others. Poor self-perception is usually the root cause of developing a "blind spot." For example, a team of people may all agree their leader is arrogant, and yet the leader may not realize his or her actions have created that impression.

The private arena consists of those personality features that are known to you but unknown to others. Because of good self-perception, these features are not blind spots. But, because of either poor communication or an unwillingness to share information with others, these features remain hidden to others. For example, a cadet may know he or she is affected by a tough home life that includes drugs and violence, but chooses to keep that information private.

The unknown arena consists of everything in your personality that is unknown to you and unknown to others. Poor self-perception may be one reason for a personality feature to remain unknown. Another reason is that personality is like an iceberg: we see a chunk of it rising above the water, but there's so much more under the surface. Unconscious and subconscious thoughts, which are normal aspects of the human mind, inhabit the unknown arena. (See page 96 for more about the unconscious mind.)

> **"The Johari Window is a tool for exploring issues of self-perception."**

THE ADJECTIVES GAME

How do you know which features of your personality inhabit each arena? The Johari Window makes use of a method that might be called "the adjectives game." Working from a fixed list of adjectives, first you select a handful of adjectives you believe describe your personality. Next, your fellow cadets, family members, friends, etc., consider that same fixed list of adjectives and choose a handful that they see in you. Finally, the selections are compared. Are you really "inspiring" the way you think you are? Or, why does everyone think you're "silly," when you just know you're an incredibly serious person? Through this process, an individual can determine what personality features inhabit their public, blind, and private arenas.[36]

athletic careful studious timid loud dishonest caring helpful outgoing irresponsible patient

THE LESSONS OF THE JOHARI WINDOW

One lesson of the Johari Window is that "feedback is the breakfast of champions." By paying attention to what people tell you in words and non-verbal cues about your behavior, you can avoid blind spots that hinder your success. Feedback allows you to expand what you know about yourself.[37]

Second, the better other people know you, the easier it is for them to work with you and support you.[38] You do this through self-disclosure, the process of telling other people things about yourself that they did not know.

At first glance, self-disclosure may seem to require you to verbally reveal secrets about yourself and be "touchy-feely." However, leaders tell people about themselves not only verbally but through their behavior. (Once again, we're reminded to lead by example.) A leader's actions, visible attitude, gestures, tone of voice, and the like are non-verbal means of self-disclosure.

The most successful teams are comprised of individuals whose healthy attitude toward self-disclosure has allowed them to become better known to their teammates.[39]

THE MYERS-BRIGGS TYPE INDICATOR

OBJECTIVES:

11. Identify the four dimensions of type, according to MBTI.
12. Describe the eight preferences of type, according to MBTI.
13. Explain why MBTI is relevant to leaders.

Personality may appear to be a vague, shapeless, squishy concept. To help people understand personality and talk about it intelligently, two researchers, Isabel Briggs Myers and her daughter, Katherine Cook Briggs, tried to bring order to the chaos by creating a model known as the Myers-Briggs Type Indicator.

According to the MBTI, there are sixteen personality types. Every person will fit into one. Which of the sixteen types is the best? None is. Personality type, according to Myers and Briggs, does not tell us who will be smart and who will be dumb, who will be charming and who will turn people off, who will find success in life and who will struggle all their days. Rather, the **MBTI merely attempts to describe our different flavors of personality.**[40]

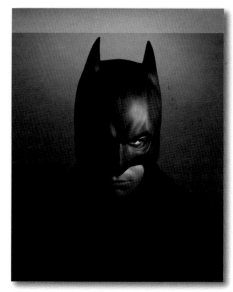

The character Batman lives squarely within the Johari window's private arena. Bruce Wayne knows who he is, but many of his friends do not.

"It's not who I am underneath, but what I do that defines me."

BRUCE WAYNE
in *Batman Begins*

On Getting Lost in the Johari Window

"No man, for any considerable period, can wear one face to himself and another to the multitude, without finally getting bewildered as to which one is true."

NATHANIEL HAWTHORNE

"MBTI helps people understand personality and talk about it intelligently."

FOUR DIMENSIONS OF TYPE

To classify each individual's personality, the MBTI looks at four aspects or dimensions of personality. Each dimension is like an old-fashioned set of scales, with one personality preference on either side. Each person's personality will naturally rest on one side of the line or the other – sometimes on an extreme end of the line, or sometimes towards the middle. For example, one dimension asks, How do you make decisions? On one side of the dimension, we have thinking, on the other side we have feeling. If you're a very logical person, you're on the thinking or T side. If you make decisions more with your heart, you're on the feeling or F side. Let's look at each dimension of type.[41]

1. Extroversion vs. Introversion
Where do you get your energy?

If you are extroverted, you prefer the world around you to the world within you. ***Extroverts enjoy spending time with people.*** The more interaction, the better. For example, extroverted students will study hard, but they prefer to study in groups. And rather than learning quietly on their own (such as through lectures), an extrovert would prefer to learn by doing some kind of group project. By being in the midst of all the action, extroverts get charged up. Because they are so outward focused, they often talk to think. If you don't know what an extrovert is thinking, you haven't been listening!

If you are introverted, you prefer the world within you to the hustle and bustle of the world around you. ***Introverts prefer to direct their energy to ideas, their imagination, and their own inner thoughts.*** Because they want to understand the world before they experience it, a lot of their world is mental. Returning to the arena of study habits, an introvert will be apt to prefer a quiet environment that allows them to concentrate fully. A common misconception about introverts is that they don't really like people. That is not necessarily true. However, after an active day around people, the introvert will look forward to time alone so they can recharge. An introvert will tell you what they are thinking, but you'll have to ask and you'll have to give them a moment to consider how they will answer you.

E - Extroverts	I – Introverts
Act, then think	Think, then act
Get energy from being around people	Get energy from spending time alone
Like to be in the midst the action	Avoid being the center of attention
Talk to think	Listen more than talk
Vocal	Quiet

The MBTI looks at four dimensions of personality. After taking a written test, each individual's personality within that dimension is "scored" or "weighed."

As seen below, people can find themselves at one extreme of the dimension or the other. You can be extremely outgoing and extroverted, for example, or only slightly so.

Keep in mind that there is no "correct" place to be on scale – MBTI merely describes your personality, it does not judge you as a person.

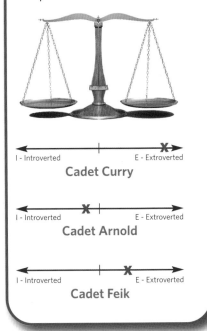

I - Introverted   E - Extroverted
Cadet Curry

I - Introverted   E - Extroverted
Cadet Arnold

I - Introverted   E - Extroverted
Cadet Feik

2. Intuition vs. Sensing
What kind of information do you naturally notice?

If you are a sensor, you are constantly taking in information using your five senses. For you, the world is real and concrete, here and now. *Sensors are interested in the specifics, the details.* They are known for having all the facts because they are literal-minded people who can recall information, especially if it is presented in black and white terms. In a cadet or military environment, the sensor will want to follow rules and regulations to the letter. A student who is a sensor is apt to prefer math and science because quite often there is clearly a "right" and "wrong" answer in those subjects.

If you are an intuitive, you take-in information by reading between the lines, looking for the "real" meaning that is not readily apparent to those looking only at surface level. Rather than focusing on the specifics or details of a situation, *intuitives look at the big picture.* The more abstract the concept, the better. In a cadet or military environment, intuitives will try to place rules and regulations in context and follow them in spirit, if not always literally. A student who is an intuitive is apt to prefer English and history because in those classes, the answers are open to discussion and debate.

S - Sensors	N - Intuitives*
Trust what is certain and concrete	Trust their own thoughts and power of deduction
Practical and want to know how new ideas will apply to them in their daily life	Imaginative and enjoy thinking about and talking about abstract ideas
Value realism and common sense	Value innovation, creativity, and "the pie in the sky"
Are specific and literal – the world is black and white	Are general and figurative – the world is a million shades of gray
Are grounded in the present	Are orientated toward the future

* Note that the letter "I" is already in use to represent introversion. Therefore, the letter "N" represents intuitives.

3. Thinking vs. Feeling
How do you make decisions?

If you are a thinker, logic, reason, and sober analysis are what you bring to the problem-solving process. *Thinkers are cool and collected, and seldom swayed by emotion.* To persuade a thinker to go along with your idea, you have to present your argument objectively and show that your plan is logical. Some may believe thinkers are heartless and uncaring. Not so. However, in the thinker's mind, emotions should not be permitted to interfere with sound judgment. Suppose a homesick cadet wanted to leave encampment on the first

---

"Are you ready to take a study break?"

SPs ask themselves, do I feel like stretching now?

SJs ask themselves, is it time for a break, what does the schedule say?

NTs ask themselves, how will taking a break now affect the rest of my day?

NFs ask themselves, if I take a break now, will I feel better?

"Can Cadet Curry come with us to the airshow?"

SPs ask themselves, do I want her to join us now?

SJs ask themselves, does the regulation permit her to join us?

NTs ask themselves, how will her participation affect the trip?

NFs ask themselves, will her participation bring feelings of goodwill?

day. In counseling the cadet, a thinker might emphasize the logical consequences that would result from leaving early (no ribbon, no Mitchell Award, lost tuition, etc.).

If you are a feeler, you bring your heart to the problem-solving process. Feelers are careful to resolve problems in a way that is consistent with their understanding of basic fairness, right and wrong. Feelers do not see problems; they see people. *If you are a feeler, you have a strong sense of empathy and prefer to consider a problem from the other person's perspective.* Some may believe feelers are irrational, not serious, and emotionally soft. Not so. Feelers can think analytically, but they prefer to trust their heart in helping them make a decision that is "right," if not perfectly "correct" or consistent. Returning to example of the homesick cadet, a feeler would try to empathize with the cadet, and highlight the potential for fun and camaraderie at encampment.

T - Thinkers	F - Feelers
Value logic, objectivity, reason	Value empathy, compassion, harmony
Analyze situations and apply rules	Empathize and find exceptions to the rules
May appear heartless	May appear soft and sentimental
Consider it more important to be truthful than tactful	Consider it more important to be tactful than truthful
Motivated by a desire for achievement	Motivated by a desire to be appreciated

4. Judging vs. Perceiving
How do you organize your life?

If you're a judger, life has a timetable. *You value structure, order, predictability.* The judger's desk is usually kept neat because every-thing has its place. When wrestling with a decision, the judger will push for closure – they prefer that matters be settled and dislike issues that remain in limbo. Constant list-makers, judgers are always trying to cross tasks off their lists so they can move on to the next scheduled project. It can be said only half-jokingly that a judger can indeed enjoy a spontaneous weekend, but only if the spontaneity has been scheduled in advance! Some fault judgers for being too goal-oriented, regimented, and resistant to change. A judger may reply that to get the most out of life you need a plan. If leading cadets on a tour of an Air Force base, a judger would be very careful to stay on schedule, observe all the rules and regulations, and not waste time dilly-dallying around the base.

THE ULTIMATE *"F"*

Agnes Gonxha Bojaxhiu, better known as Mother Teresa or "The Saint of the Gutters," clothed, fed, and healed thou-sands of India's poor, people so destitute society pronounced them "untouchable."

Calling on the people of the world to follow her example, she once said, "If you can't feed a hundred people, then just feed one."[42]

From a rationalist perspec-tive, that makes little sense. Feed just one? That would not solve the problem of global hunger. But Mother Teresa, a feeler, knew in her heart it would make a difference to that one receiving help.

Likewise, she once mar-veled, "I will never understand all the good that a simple smile can accomplish."[43]

A leader admired more for her heart than her logic, Mother Teresa was awarded the Nobel Peace Prize in 1979.

If you are a perceiver, flexibility and spontaneity are your favorite words. *If you are a perceiver, you prefer to keep your options open.* You dislike feeling regimented or forced to live by a timetable, even one that you created. Because perceivers are so flexible, they are able to adapt to changing situations and may even enjoy a chaotic environment. Look at the perceiver's desk, and you're apt to find it cluttered. Some fault perceivers for not being goal orientated, for being procrastinators, and for having a wandering mind. In their defense, perceivers might reply that too much structure limits their ability to explore life's many possibilities. If leading the base tour mentioned above, a perceiver would keep their options open. They would allow successful tours to run over schedule. If exotic aircraft unexpectedly appear in the pattern, the perceiver will make time for the cadets to watch.

J - Judgers	P - Perceivers
Value structure, order, and regimentation	Value flexibility and spontaneity
Set goals and proceed to fulfill them methodically	May set vague goals and change them as new opportunities arise
Feel good when finishing a project and crossing it off their list	Feel good when starting a new project and sensing its many new possibilities
Push for closure and are most comfortable only a after a decision has been reached	Prefer not to feel pressured into making decision and are most comfortable when their options remain open
Schedule time in their day to have fun	Allow time for fun even when others worry about deadlines

THE ULTIMATE "T"

"Live long and prosper" is his way of saying, "have a nice day."

Spock, from *Star Trek*, is one of the most logical, rational, unemotional characters ever created. In MBTI parlance, he is a thinker, a "T" to the extreme.

When he sees something he likes, Spock says, "Fascinating," indicating his intellectual interest in the subject.

Spock tries hard to hide his emotions. While others aboard the *Enterprise* let their feelings guide their decisions, Spock prides himself on his ultra-cool logic.

"You'd make a very splendid computer," observed Captain Kirk, to which Spock replied, "That is very kind of you, captain."[44]

A SYSTEM DESCRIBING SIXTEEN PERSONALITY PREFERENCES

With four dimensions having two preferences each, the math works out such that there are sixteen combinations possible in the MBTI. This table shows the sixteen personality types, along with slogans that describe them.[45]

THE MBTI IN CONTEXT FOR LEADERS

Cadets should not look at MBTI as the definitive statement about their personality. Each of us is more than what a 4-letter code represents. However, MBTI offers a useful way to talk

ISTJ	ISFJ	INFJ	INTJ
"Doing what should be done"	"A high sense of duty"	"An inspiration to others"	"Just ask and I can improve anything"
ISTP	ISFP	INFP	INTP
"Ready to try anyting once"	"Sees much but shares little"	"Quietly and nobly serving society"	"A love of problem solving"
ESTP	ESFP	ENFP	ENTP
"The ultimate realists"	"You only go around once in life"	"Giving life an extra squeeze"	"One exciting challenge after another"
ESTJ	ESFJ	ENFJ	ENTJ
"Happiness is following the rules"	"Hosts and hostesses of the world"	"The public relations specialists"	"Everything's fine: I'm in charge"

about personality. Its system of preferences helps explain why people favor certain approaches to life. *The MBTI supports a leader's development by helping the leader understand themselves so that they may be more effective in working with others.* Cadets in particular may find MBTI helpful as they search for careers that match their personality.[46]

MOTIVATION

In chapter four, we introduced the concept of motivation and defined it as the reason for an action, the "why" that causes someone to do something.

Building on that understanding, next we will consider what actually motivates people. Is it cold, hard cash? The warm feeling that comes from helping others? An electric shock? Rank and prestige? An understanding of what motivates people is essential knowledge for leaders.

MASLOW'S HIERARCHY OF NEEDS

OBJECTIVES:
14. Describe each of the five basic needs, according to Maslow.
15. Explain why Maslow uses a pyramid to describe his model.
16. Suggest ways a leader can help individuals fulfill each of the five basic needs.

Nothing else matters if you have a toothache. And for the world's poorest, the worry of finding a meal is the only thing on their mind.

Abraham Maslow was a psychologist who wanted to find a comprehensive scheme to explain what motivates people. ***His basic theory says individuals are motivated by unfulfilled needs.*** According to Maslow, there are five basic needs and they are arranged in the shape of a pyramid. Human beings focus on the lowest unfulfilled need. Once that need is met, higher needs emerge, and people work their way up the pyramid.[47] What follows is a description of each of the five needs in Maslow's hierarchy.[48]

> **"According to Maslow, humans focus on the lowest unfulfilled need on his pyramid."**

PHYSIOLOGICAL NEEDS

A human being who is lacking everything in life will hunger most for food and water. Although this individual is lacking love, an education, success, or a trophy, it will be food and water that occupy his mind. Physiological needs also refer to the basic operation of the human body. America's first astronaut, Alan Shepard, began his mission with total focus. After numerous delays, all he could think about was his need to use the restroom. NASA engineers famously concluded they had to grant him permission to go in his spacesuit.[49] ***Unless the physiological needs are met, all other needs are forgotten or even denied.***

HE HAD TO GO

According to Maslow's hierarchy of needs, a person will focus on their lowest unfulfilled need.

In other words, you cannot expect to launch into space and achieve self-actualization if, like America's first astronaut Alan Shepard, all you can think about is your dire urge to use the restroom.

Although somewhat crude, this story illustrates an important principle. Leaders must help their teammates fulfill the lowest needs before the team will be motivated to reach its full potential.

SAFETY NEEDS

Once the physiological needs are satisfied, the need for safety emerges. *Safety includes freedom from fear, violence, and uncertainty.* The loud scary noises of a thunderstorm will cause a baby to cry out, illustrating the baby's fear and need for safety. Schedules, routines, and a familiar and comfortable home satisfy our need for safety, too. America's stable government and well-functioning society, where crime is minimal and no dictator can knock on our door in the night, also help satisfy our safety needs.

Self - Actualization Needs

Esteem Needs

Love or Belonging Needs

Safety Needs

Physiological Needs

Diagram of Maslow's Hierarchy of Needs

LOVE OR BELONGING NEEDS

"Take away love," wrote the poet Robert Browning, "and our earth is a tomb."[50] After achieving relative safety, Maslow argues that *we are motivated by a need for love or a basic connection with other people, a sense of belonging.* As a baby, you felt this need whenever mom was not close. The feelings we have for friends, family, boyfriends, girlfriends or spouses are expressions of the love or belonging needs. One reason people choose to join clubs and teams is to satisfy their social needs, their need to belong.

ESTEEM NEEDS

"I'm not the greatest, I'm the double greatest," proclaimed boxer Muhammad Ali.[51] Clearly, Ali was a man who possessed self-confidence. Maslow's fourth need cuts two ways. First, the need for esteem focuses inward. Everyone wants to feel good about themselves. Second is an outward need for esteem. *We are motivated by a desire for attention, honor, appreciation, and a good reputation.* The esteem need is what motivates actors to strive for the Oscar, for cadets to compete for the Honor Cadet trophy.

Camaraderie
One reason people join clubs and civic groups is to satisfy their natural need for belonging. Therefore, in a volunteer organization like CAP, it's especially important for leaders to allow cadets to have fun and make friends.

SELF-ACTUALIZATION NEEDS

At the pinnacle of Maslow's pyramid is the need for self-actualization, the desire for self-fulfillment. In describing this need, Maslow said, *"What a man can be, he must be."*[52] People who achieve self-actualization have truly lived up to their potential. Of all the wonderful, famous works of art he created, Michelangelo signed only one, the *Pieta*.[53] Maslow would say the *Pieta* represents Michelangelo finally achieving self-actualization. While you don't need to be the world's greatest at something to achieve self-actualization, Maslow believed that very few people fulfill the highest need on his pyramid because in modern society, basically satisfied people are the exception. Moreover, an individual may find self-actualization in one aspect of life but struggle to fulfill basic needs in some other area of their life.

MASLOW AND THE LEADER

What does Maslow's hierarchy offer the leader? ***The hierarchy of needs gives leaders a framework for understanding what motivates people.*** The pyramid shape reminds leaders that certain motivations are stronger than others and must be satisfied before the higher needs emerge. Maslow also shows us that while everyone is different, everyone shares the same basic needs.

HAWTHORNE STUDIES

> **"The discovery of human beings in the workplace."**

OBJECTIVE:

17. Identify the key lesson of the Hawthorne studies.

A century or more ago, managers thought workers were like machines. They tried to use scientific principles to make the employees more productive, much like a 21st century technician might calibrate a robot.

In what became known as the Hawthorne studies, researchers tried to boost productivity in a factory by adjusting the lighting, rearranging the times when employees took their breaks, and tweaking the employees' work schedules. All these changes began as attempts to fine tune the work environment and find the best conditions to manufacture their products.[54]

SURPRISING RESULTS

As the researchers made their adjustments, productivity increased. Was it really the fact that 100-watt light bulbs had replaced 60-watt light bulbs? Was high productivity a result of taking a break at 10 am instead of 10:30?

To test their hypothesis, the researchers returned the environment to normal. They plugged the old light bulbs back in and sent the workers to their coffee breaks according to the original schedule. To everyone's surprise, productivity kept climbing. Why?

A LESSON FOR LEADERS

The lesson of the Hawthorne studies is that when leaders pay attention to their people and treat them as partners, people feel appreciated and will perform better.[55] Because management was actually talking with the workers at the Hawthorne plant, the workers felt important. Instead of working only hard enough to not get fired, the employees were true collaborators. As such, they became more self-motivated and took greater interest in their jobs.

Good leadership is what motivated the employees and boosted productivity, not the light bulbs. One scholar called the Hawthorne studies, "the discovery of human beings in the workplace."[56]

Hawthorne in Action
Think of your people as collaborators. That's the overall lesson of the Hawthorne studies. The Cadet Advisory Council is a good example of that principle in action. CAC representatives are consulted about major issues facing CAP.

CLASSICAL CONDITIONING

If you have a dog, perhaps he comes running if he hears you open the cookie jar. Jingle his leash and Chewy knows it's time for a walk. We say our dogs are smart, but really they have simply learned to associate sights and sounds with certain actions.

PAVLOV'S DOGS

Ivan Pavlov studied the digestive system in dogs. He learned that dogs produced saliva to help them chew and swallow their food. Pavlov's breakthrough came when he noticed his dogs drooling even though no food was around. Why? He discovered that the dogs were reacting to his lab coat. Whenever Pavlov had fed his dogs, he was wearing a white lab coat. Therefore, show the dogs a lab coat and they'll assume it's time for supper.

Time to Play
Grab his leash and Chewy will know it's time for a walk. But how does he know? Classical conditioning. He's learned to connect a stimulus (sight and sound of the leash) to a reflex (run to the door and get ready to go outside).

Realizing he was on to something, Pavlov conducted an experiment. Moments before feeding his dogs, Pavlov rang a bell. If the bell sounded in close association with their feeding, the dogs learned to associate the sound with food. Eventually, Pavlov could make his dogs drool simply by ringing a bell. (As the old joke goes, Pavlov: Does that name ring a bell?)

"Pavlov learned that a stimulus can trigger a reflex."

Pavlov's discovery was that environmental events (the things going on around the dog) that previously had no connection to a given stimulus (such as the sound of a bell ringing) could trigger a reflex (drooling). In short, ***Pavlov discovered what is called classical conditioning, the process whereby a living thing (e.g.: a person) learns to connect a stimulus to a reflex.***[57]

IMPLICATIONS FOR LEADERS

What does Pavlov's discovery add to our understanding of leadership? If a leader praises people for a job well done, they are apt to continue working hard. The praise is said to ***reinforce*** the desired behavior. But there is a downside to this style of leadership. It succeeds only when the individual is prepared to work for the reward offered.[58] For example, when you were little, perhaps you could be motivated by the promise of a cookie, but now in your teens, a cookie no longer motivates. At some point, praise becomes routine and fails to truly motivate.

Although Pavlov proved he could train his dogs to drool by ringing a bell, *leadership experts believe classical conditioning is too simplistic a method to motivate people to pursue a goal.*[59] It is the "carrot and stick" way of leading. Eventually, the leader will run out of carrots and/or the follower will lose his appetite for carrots and hunger for something more.

As discussed earlier, today's leaders motivate by aligning personal goals with team goals. Or they may appeal to their followers' desire for belonging, esteem, or self-actualization, as suggested by Maslow. There are several other more sophisticated ways to motivate, which we will discuss in later chapters. Pavlov's bell worked on dogs. It is a mistake for leaders to believe they can use the techniques of dog obedience training to effectively motivate intelligent, complex people.

CLASSICAL CONDITIONING: FOUR WAYS TO CHANGE BEHAVIOR

Building on Pavlov's discoveries, psychologists learned how to change behavior through more sophisticated ways of conditioning. Here's a look at four approaches.[60]

Positive Reenforcement happens when a pleasant reward is used to increase the frequency of a behavior.

Do X ⟶ Y happens ⟶ Feel good ⟶ Do X again

A student brings home straight A's. His parents give him $100. The student learns that good grades pay.

In a store, a child whines for mom to buy her a toy. To quiet the child, the mom buys her the toy. The child learns that if she whines, she gets what she wants. Unfortunately, mom reinforced the wrong behavior.

Negative Reenforcement happens when an unpleasant stimulus is removed to encourage the desired behavior.

Do Z ⟶ Feel bad

Do X ⟶ Feel better ⟶ Do X again

A teen experiments with marijuana. The dope makes him sick and he gets in trouble. Instead of hanging out with trouble-making druggies, the teen joins CAP, participates in lots of great activities, and feels better. Wanting to keep feeling better, the cadet becomes even more active in CAP.

Punishment is any stimulus that represses or stops a behavior.

Do A ⟶ Z happens ⟶ Feel bad ⟶ Do A less

A teen texts his girlfriend dozens of times a day for a month. When the phone bill arrives, his parents are furious and he is grounded. He learns to not text his girlfriend so much.

Extinction refers to the reduction of some response that the person had previously displayed.

Want Y ⟶ Do X ⟶ Don't get Y ⟶ Stop doing X

A boy wants a girl to pay attention to him. He tells her dirty jokes. She's not impressed and walks away. The boy learns that telling dirty jokes is not the way into a girl's heart.

THE MILGRAM EXPERIMENT

OBJECTIVES:

26. Summarize the events of the Milgram experiment.
27. Identify the key lesson learned from the Milgram experiment.
28. Explain why the Milgram experiment is a warning for leaders.

How much pain will an ordinary person inflict on another, simply because they are told to do so? Welcome to the Milgram experiment.[61]

You and another volunteer, a stranger, enter a laboratory at a prestigious university. You are seated before a control panel with dials and switches. The other volunteer takes a seat in a nearby room. Wires are connected to his body, but then a curtain is drawn. You'll be able to hear this other volunteer, but can no longer see him. Your host, a researcher wearing a white lab coat, explains he wants to conduct an experiment involving a study of memory and punishment.

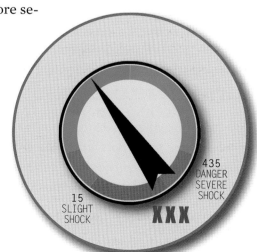

The Researcher

The Teacher
*The experiment's **real** subject*

The Learner
Secretly is a friend of the researcher

Diagram of the Milgram Experiment

RULES OF THE GAME

One person, the "learner," will be playing a simple memory game, testing his ability to recall words in a certain order.

Another person, the "teacher," will test the student's ability to learn by administering an electric shock to the learner whenever the learner gets a question wrong.

You will act as the "teacher," and the second volunteer, whom you do not know, will be the "learner."

> **"If you were ordered to, would you torture someone?"**

If the learner answers a question correctly, he will receive no punishment. But, if he answers incorrectly, you will administer an electric shock to punish him. The shocks will become more severe with every incorrect answer.

THE APPARATUS

Looking at the electrical apparatus that controls the shocks, you see a dial that has settings ranging from 15 to 450. The "15" setting is labeled as "Slight Shock" and the "435" is labeled as "Danger Severe Shock," and the "450" is simply labeled "XXX."

HOW THE EXPERIMENT PROCEEDED

Ready to begin? Blue, sky, dog, cat... the learner tries to remember sequences of simple words like this. Sometimes he recalls the words correctly, sometimes not. Wrong answer: administer a shock.

When the shock dial reaches 75, the learner grunts audibly.

At 120, the learner says the shock is becoming painful. He's squirming in his seat. Sweat pours down his face.

"Please continue," the researcher instructs you.

At 150, he screams, "Get me out of here, I refuse to go on!!"

"The experiment requires you to continue," states the researcher.

At 180, the learner calls out, "I can't stand the pain!!" He's screaming louder, flailing his arms, in visible agony.

At 270, he howls. Literally. The man is in such pain, he howls.

"You have no other choice," the researcher informs you, "you must go on."

At 330, there is only silence.

"You must continue..."[62]

Milgram's Apparatus
The "teacher" believed this machine really could deliver electric shocks to the "learner," but Milgram's apparatus was fake.

THE RESULTS

How severely would you be willing to shock the learner, a complete stranger? What percent of people would be willing to turn the shock dial all the way up, to 430, and then beyond, to its maximum at "XXX"?

Researcher Stanley Milgram found that 63 percent of the men who acted as "teachers" were willing to carry this diabolical experiment to the end, turning the dial all the way up to "XXX."[63]

> "Milgram discovered that obedience to authority is a powerful motivator."

Fortunately, the electrical shock apparatus was fake. The "learner" was an actor. Only the individuals serving as "teachers" were unaware of what was really going on, that they were the real subjects of this experiment.

WHAT MILGRAM LEARNED

What does the Milgram experiment teach us? *Milgram discovered that obedience to authority is a powerful motivator.* Ordinary people, it turns out, can commit terrible atrocities simply because they feel duty-bound to obey authority figures.[64]

Believing that the "learners" were experiencing tremendous pain, the "teachers" would sweat, tremble, stutter, groan, dig their fingernails into their own flesh. Nevertheless, nearly 2 in 3 people complied to the researcher's simple commands to continue. They were willing to torture another person, simply because a scientist in a lab coat told them to.

LESSONS FOR LEADERS

"Unthinking respect for authority," in the words of Einstein, "is the greatest enemy of truth."[65] The Milgram experiment reminds leaders that blind obedience is not real obedience. In the military, leaders study the concept of "lawful orders," and learn that disobedience to those in authority can in fact be obedience to the higher obligations of our democratic ideals and Core Values.[66] Moreover, *Milgram shows that leaders, in light of their power over their followers, bear some responsibility for the actions of their subordinates.* "The disappearance of a sense of responsibility," observed Milgram, "is the most far-reaching consequence of submission to authority."[67]

INTER-PERSONAL CONFLICT

In any relationship there will be conflict. "Conflict is inevitable," observes one leadership expert, "but combat is optional."[68] The real measure of a leader then is how he or she handles conflict.

Even the strongest relationships will experience bumps along the way. It is naive to believe otherwise. At one time or another, interpersonal conflict will be present on every team, in every friendship, within every family. Nations rise and fall depending on whether their political system is equipped to resolve conflict fairly.

Therefore, leaders try to understand what can give rise to conflict so that they might steer clear of it, when possible. And leaders work to develop skills enabling them to manage conflict in a productive, ethical way.

Supreme Conflict
What's the ultimate method for resolving conflict in the United States? Taking the case to the Supreme Court. Here, cadets pose with Associate Justice Antonin Scalia. Nations rise and fall depending on whether their political system is equipped to resolve conflict fairly.

DEFENSE MECHANISMS

OBJECTIVES:

29.   Define the term, "defense mechanism."

30.   Explain why people naturally turn to defensive behavior.

31.   Give examples of how some defense mechanisms play out in everyday life.

32.   Defend the idea that a basic knowledge of defensive behavior is relevant to leaders.

What happens when someone cannot handle anxiety, stress, and pressure? What happens when the burdens of dealing with other people and our own feelings of guilt and failure become overwhelming? ***Defense mechanisms activate to protect us from psychological injury.***[69] People turn to their defense mechanisms especially when their sense of self worth is challenged by their own inner feelings or by the actions of other people.[70] In short, ***defense mechanisms are behaviors people use to deal with anxiety, stress, or pressure.***[71]

Defensive behavior is perfectly normal. It is a natural and often unconscious reaction to emotional pain.[72] When you were little, perhaps you were incredibly shy. Given the opportunity to meet someone new, you might have run and hid behind your mother. If making a big speech in front of a large group for the first time, you might try to use humor as a crutch to hide your nervousness. Because everyone experiences anxiety, stress, and life's pressures, everyone turns to their defense mechanisms from time to time, whether consciously or unconsciously. However, most people who are emotionally healthy come to a point where they face their problems and learn to rely less on their defense mechanisms.[73]

Although defensive behavior is a natural reaction to emotional pain, it can become a problem because it changes the way we see reality.[74] It makes it hard for us to be honest with ourselves. No one enjoys holding a mirror up to themselves and looking for shortcomings. It is especially difficult to overcome a natural tendency to use defense mechanisms because they are often unconscious reactions.[75] After all, how do you learn about something you do not realize you are doing?

More specifically, defensive behavior can become habit forming. Just as a smoker is always used to having a cigarette between their fingers, it is possible for individuals to find themselves constantly relying on defensive behavior to deal with stress. In such cases, defensive behavior becomes unhealthy when it stops us from ever facing problems head-on. Moreover, when life is spent in a defensive crouch, when someone is always turning to their defense mechanisms,

> **"Defense mechanisms are behaviors people use to deal with stress."**

The Unconscious Mind
Defense mechanisms are often unconscious. The human mind is like an iceberg. We're aware of many of our own thoughts. (We're conscious of the tip of the iceberg.) But there is much more invisibly lurking under the surface of our minds. We call the hidden mind the unconscious.

little energy is left over to do what they really want to do.[76] Who can be happy and successful when their mind is perpetually trying to withstand attacks?

TYPES OF DEFENSE MECHANISMS

Because defensive behavior is so common, psychologists disagree as to exactly how many different types of defense mechanisms exist.[77] Here's a sampling of some of the most common ones:

Displacement *occurs when someone redirects feelings about something onto something less threatening.*[78] For example, imagine you are angry at your teacher. You know that if you yell at her you'll get in trouble. Instead, you displace your anger by waiting until you get home, at which time you yell at your dog and push him away when he greets you.

Projection *is the act of taking one's own unacknowledged thoughts or feelings and falsely attributing them to someone else.*[79] Put another way, instead of facing the bad feelings you have about yourself, you try to say another person is struggling with those feelings. For example, a young cadet who is away from may home for the first time may deal with her fear by picking on another cadet: "You're a big baby. You probably miss your mommy and need a night light!"

TYPES OF
DEFENSE MECHANISMS

Displacement

Projection

Rationalization

Intellectualization

Denial

Suppression

Withdrawal

Rationalization *is when someone devises reassuring or self-serving explanations for their behavior.*[80] They attempt to use fancy, twisted thinking to avoid facing a problem. It is the "sour grapes" defense. In Aesop's fables, the fox was hungry for grapes, but he couldn't reach them. "They were probably sour anyway," he rationalized, even though deep down he knew they would have been delicious.[81]

SEE NO EVIL   HEAR NO EVIL   SPEAK NO EVIL

Intellectualization is similar to rationalization in that both use some form of twisted thinking. *Through intellectualization, a person tries to remove the emotional side of a situation and instead examines their problem in an excessively abstract way.*[82] They withdraw into a scientific mindset out of fear of facing powerful emotions head-on. For example, if dumped by his girlfriend, a boy might avoid feelings of heartbreak by telling himself, "When one considers the median duration of high school romantic relationships in suburban America, one must conclude that my relationship had only a 76.4% chance of lasting through prom."

Denial, according to the old corny joke, is not just a river in Egypt. Rather, *denial is a mechanism in which a person fails to acknowl-*

Suppression
Have you ever had a problem so troubling you couldn't bear to acknowledge it existed? You refuse to let yourself see it, hear it, or speak of it. That's suppression.

edge facts that would be apparent to others.[83] When someone refuses to acknowledge what has or will happen, they are in denial. A cadet who is 80 pounds overweight, for example, might deny that he is in fact obese. "I'm a little chubby," he might say, "but it's no big problem."

Suppression *is when a person knows they have anxieties or problems, but they set them aside, choosing not to even think about them.*[84] Suppose a boy left the restroom at school and did not realize he forgot to zip his fly. Upon returning to class, everyone notices, points, and laughs. That experience would be a good candidate for suppression. The unfortunate boy knows what happened, but prefers never to think about it again.

Withdrawing to a quiet spot can be a good way to cope with stress. But if this girl does not resume her normal activities soon, her withdrawal could become a problem.

Withdrawal *entails removing oneself from events, people, things, etc., that bring to mind painful thoughts and feelings.*[85] For example, rather than face the fear of girls rejecting them at a school dance, a group of boys may withdraw to a corner of the room, hoping everyone forgets they are even present. Withdrawal can be a cause of loneliness and alienation, which in turn, brings about even more problems. In extreme cases, it can lead to alcohol or drug abuse.[86]

DEFENSIVE BEHAVIOR AND THE LEADER

Why is an understanding of defense mechanisms important to leaders? Defense mechanisms mask problems; leaders help people overcome problems. ***Knowing something about defensive behavior better enables the leader to spot anxiety, stress, and pressure among followers.*** Moreover, this understanding can help leaders become more self-aware of their own individual feelings of guilt, frustration, or failure that are visible to others but invisible to themselves.

Withdrawal

Displacement

This athlete is taking his anger out on the speedbag. Feeling defensive is natural in times of stress. At least this man is channeling his stress in a healthy way.

Uganda's dictator, Idi Amin, awarded himself countless medals, badges, and sashes. Were those decorations his defense mechanism against a poor self-image?

Projection

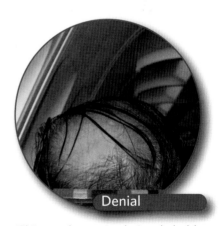
Denial

This gentleman is obviously bald. The comb-over isn't fooling anyone, except himself.

THE INEVITABILITY OF CONFLICT

OBJECTIVES:

33. Define "conflict."
34. Compare the people skills of "knee-jerk" and "deliberate" leaders.

> *"Man is harder than rock and more fragile than an egg."*
> - Serbian proverb

Where there are people, there will be problems. Conflict is unavoidable. ***Managing conflict is a normal and inevitable part of leadership.***[87]

What is conflict? It is more than a mere difference of opinion. ***Conflict is a disagreement through which individuals perceive a threat to their needs, interests, or concerns.***[88] It develops when someone does not act as another wants. The most common conflicts cadet NCOs will be called upon to help resolve are personality conflicts, which are especially irrational. The individuals' perceptions and emotions take hold at the expense of logic and the sober grasp of reality.[89]

PEOPLE SKILLS

According to one definition, leadership is the deliberate process of sharing values. That means leadership is supposed to be a conscious, intentional act. Leaders sometimes run into problems because they act without really thinking. This chart illustrates how our natural reactions are not always professional, and therefore can create conflict.[90] Are you a knee-jerk or deliberate leader?

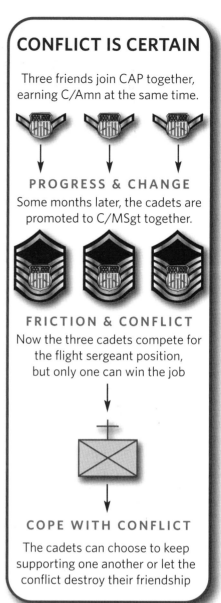

CONFLICT IS CERTAIN

Three friends join CAP together, earning C/Amn at the same time.

PROGRESS & CHANGE
Some months later, the cadets are promoted to C/MSgt together.

FRICTION & CONFLICT
Now the three cadets compete for the flight sergeant position, but only one can win the job

COPE WITH CONFLICT
The cadets can choose to keep supporting one another or let the conflict destroy their friendship

Natural Reactions of the Knee-Jerk Leader

1. Reacts emotionally to bad news

2. Fights and acts defensively in the face of threats

3. Makes snap judgments about people, ideas, and events

4. Focuses on status, rank, and awards

5. Dwells on past successes

6. Relies on rank to get the job done, the "I order you to..." approach

7. Suffers from the "not invented here" syndrome

8. Displays the "we've always done it that way" syndrome

9. Runs to the boss to settle disputes

10. Gossips about other people, especially around subordinates

The Professionalism of the Deliberate Leader

1. Steps back, thinks, remains calm

2. Monitors own attitudes and defensive behaviors

3. Gives people the benefit of the doubt and carefully considers new ideas

4. Concentrates on doing a good job and trusts that awards will follow

5. Focuses on the future

6. Applies leadership skills to get the job done

7. Seeks good ideas no matter where they come from

8. Values innovation and creativity

9. Settles disputes without running to "mom"

10. Shows commitment to Core Values by refraining from gossip

THE LEADER'S ROLE IN MANAGING CONFLICT

OBJECTIVE:
35.   Defend the idea that leaders have a role in managing conflict.

Leaders are interested in managing conflict because conflict destroys teamwork and therefore limits the team's ability to succeed. Our study of Maslow's hierarchy of needs illustrates that when two people fight, they are not satisfying all their needs. Leaders step in to resolve conflict so that the warring individuals might return to a path leading to self-actualization.

Moreover, ***conflict often creates an inhospitable feeling that can affect everyone on the team***, even bystanders who are not directly involved. This is especially true when two leaders develop a conflict – their ill feelings are bound to poison their teams.[91]

POSSIBLE OUTCOMES

Think about what can result from conflict and you'll see that there are four possible outcomes:

WIN / WIN   WIN / LOSE   LOSE / WIN   LOSE / LOSE

Although logic tells us this is true, leadership experts believe that ***unresolved conflict is itself the enemy.***[92] One expert observes, "There won't be any losers unless we all lose, and there won't be any winners unless we all win."[93] Former president Jimmy Carter once said, "Unless both sides win, no agreement can be permanent."[94]

THE DUTY TO INTERVENE

In a formal organization like CAP, the military, a school, or business, leaders may have a legal obligation to manage conflict that arises among their subordinates. Conflict can take the form of sexual harassment, racial bigotry, or threats of violence. Democratic societies expect that those who have authority will discipline followers who break the rules. In short, ***the leader has a duty to respond to conflict because the leader is responsible for the team's behavior and its success.***

EAST & WEST GERMANY

After World War II, Germany was divided into East Germany, controlled by the Soviet Union, and West Germany, controlled by the U.S. and its allies.

As the people of the West grew into a prosperous democracy, the government of the East built walls to keep their people from leaving.

Fortunately, at the end of the Cold War in 1989, the East allowed protesters to tear down the Berlin Wall. Today, Germany is united as a single nation and boasts one of Europe's strongest economies.[96]

UNRESOLVED CONFLICT IS THE ENEMY

Conflict is inevitable and natural. The real test of a leader is how he or she deals with that conflict. Therefore, one expert observed, "unresolved conflict is the enemy." The stories of Korea and Germany illustrate that principle.

NORTH & SOUTH KOREA

In the 1950s, war ravaged the Korean peninsula. Armies in the North were supported by communists. Armies in the South were supported by the U.S. and the United Nations.

When did the war end? It hasn't! Technically, North and South Korea remain at war to this day, though both sides signed a cease fire in 1953.

Today, South Korea is a prosperous democratic nation. A dictator rules North Korea and its people are starving.[95]

North & South Korea remain divided

At night, it's easy to see the border between North & South Korea

Fall of the Berlin Wall, 1989

Berlin today, peaceful and prosperous

100

AT THE AIR FORCE ACADEMY:
THE LEADER'S DUTY TO MANAGE CONFLICT

In the 1990s, the Pentagon faulted Air Force Academy leaders for failing to see the seriousness of sexual assault and harassment at the academy.[97]

Women at the academy had reported 150 sexual assaults over 10 years. What the Pentagon's inspector general found most troubling was that commanders took little action until the accusations became public.

"We conclude that the overall root cause of sexual assault problems at the Air Force Academy was the failure of successive chains of command over the past 10 years to acknowledge the severity of the problem," wrote the IG. He added, "They failed to... change the culture until [the problem made national news]."[98]

In short, eight leaders were made to share the blame for not stepping in when they should have.

No one suggests that those eight leaders personally condone sexual harassment or assault. Nevertheless they were held responsible because the problem happened on their watch. Leaders have a duty to manage conflict.

METHODS FOR MANAGING CONFLICT

OBJECTIVES:
36. Describe five basic approaches to managing conflict.
37. Outline a process for mediating a conflict between two people.

In managing conflict, leaders focus on changing people's behavior.[99] Outward actions count for more than inward feelings. Two teammates are not required to like one another, but they must be capable of working together. A leader may hope that each subordinate has good feelings about their teammates, and indeed good feelings can only help a team, but in the end, it's the ability to work together that is the mark of true professionalism.

BASIC APPROACHES

What are some basic approaches to dealing with conflict? What follows is a brief overview:

Avoidance ***occurs when leaders recognize conflict exists but they choose not to engage the problem.***[100] In the best case, avoidance is about choosing one's battles wisely. In the worst case, avoidance will make the problem grow worse.

Denial ***is when the leader refuses to acknowledge the conflict exists.***[101] Denial represents a failure of leadership.

METHODS OF MANAGING CONFLICT
Avoidance
Denial
Suppression & Smoothing
Compromise
Zero-Sum Game
Mediation

Suppression and smoothing *is a two-pronged approach.* First the leader suppresses conflict, suggesting it is not as bad as it seems. Second, the leader smoothes over differences to suggest that both parties are aiming for the same goal. This technique may be successful for a while, but in the final analysis, it is a form of avoidance.[102]

Compromise *is an attempt to create a win/win situation.*[103] To settle their differences, each side makes concessions. Every kindergartner understands compromise. It is part of our democratic heritage. Compromise can be problematic when individuals believe core principles are involved from which they cannot retreat.

The Zero-Sum Game *sees conflict in only win/lose terms.* One party must win and the other must lose an exactly equal amount.[104] For example, two cadets share a pizza that has eight slices. From a zero-sum game perspective, any slice eaten by the first cadet represents one slice that the second cadet cannot eat. Former president Clinton observed:

> The more complex societies get and the more complex the networks of interdependence within and beyond community and national borders get, the more people are forced in their own interests to find non-zero-sum solutions. That is, win–win solutions instead of win–lose solutions.... Because we find as our interdependence increases that, on the whole, we do better when other people do better as well . . . we have to find ways that we can all win, we have to accommodate each other.[105]

Mediation *is an attempt to resolve conflict by using a third party to facilitate a decision.*[106] A mediator is like a judge. The mediator's decisions may or may not be binding. When leaders are called to mediate personality conflicts, they must maintain impartiality so that both sides accept the decision as fair. This fact underscores the importance of leaders not showing favoritism to their subordinates.

GROUND RULES FOR MEDIATING CONFLICT

1. Arrange to meet the two conflicting individuals in private, on neutral ground.

2. Allow only one person at a time to talk.

3. State that the discussion is to remain confidential. This encourages everyone to speak their mind.

4. Listen to understand. Try to identify the conflict's root cause.

5. Prohibit gossip or hearsay. Insist the discussion refer only to people who are present.

6. Focus the conflicting individuals on attacking the issues, not one another.

Concession. when one party yields a right or a benefit in hopes that the other will yield an equivalent right or benefit[107]

A PROCESS FOR MEDIATING CONFLICT

1. **Set a positive tone.** Open by explaining you are not here to judge, but to facilitate a solution. Remind the individuals that no one can win unless everyone wins, and no one loses unless we all lose.

2. **Be mindful of appearances.** Even seating arrangements can inadvertently signal that one person has the upper hand.

3. **Allow the first person to talk** and explain their side of the story, without interruption.

4. **Allow the second person to talk** and explain their side of the story, without interruption.

5. **Summarize your understanding of the conflict.** In summarizing, focus on points of professional conflict.

6. **Begin the interview stage.** Direct specific questions to each individual. Questions should be logical and push the conflicting people beyond their tired old stories. Plant seeds for a solution.

7. **Ask each person how the conflict can be resolved,** especially in light of any lessons they may have learned as a result of the discussion.

8. **Ask each individual to make concessions,** if necessary.

9. **Aim for a consensus,** a general agreement that everyone can live with. Persuade both individuals to accept the solution that seems most fair.

10. **Conclude** by asking everyone to shake hands.

LEADING IN A DIVERSE SOCIETY

OBJECTIVE:

38. Define the term "diversity" in your own words.

Having respect for diversity is a personal decision.[108] As discussed in chapter one, CAP's Core Value of respect is the price of admission into our organization. The Core Values teach us that every person is worthy of respect simply by virtue of their basic human dignity. Moreover, **as a nation of immigrants, America's diversity is its strength.** Ignorance, insensitivity, and bigotry can turn that diversity into a source of prejudice and discrimination.[109]

> **"Diversity is the greatest strength of our Air Force."**

Red-Tailed Devils
The Tuskegee Airmen were known for their aircrafts' distinctive red tails.

DIVERSITY IN THE MILITARY & CAP

OBJECTIVES:

39. Defend the claim that diversity is important to the military and CAP

Tuskegee Airmen, 1941-45
During WWII, some believed blacks lacked the intelligence, skill, and patriotism to fly combat missions. The Tuskegee Airmen paved the way for diversity as one of the most successful fighter groups of the war.

The armed forces were one of the first American institutions to racially desegregate. In 1948, President Truman ordered, "There shall be equality of treatment and opportunity for all persons [in the military] without regard to race."[110] The military broke the color barrier six years before public schools desegregated, and sixteen years before Congress made it illegal for private businesses to discriminate on the basis of sex, race, or religion. It can be said that the desegregation of the military helped launch the civil rights movement of the 1960s and that the military's success made the business sector take notice.[111] "Diversity," said a former Chief Master Sergeant of the Air Force, "is the greatest strength of our Air Force... The Air Force attracts men and women from all walks of life; we welcome these teammates and value their differences."[112]

Racial Desegregation. the overturning of laws that had required people of different races to live separately

CAP was founded upon a commitment to diversity. The Patrol welcomed people who wanted to serve America during World War II but were unqualified for military service due to age or physical disability.[113] CAP's first national commander, Maj Gen John Curry, was particularly interested in recruiting females in a day when women's opportunities were limited. Our first national commander was a progressive whose respect for diversity was ahead of its time.[114]

All-Female Crew, 2005
These six airmen flew a C-130 into combat over Afghanistan as the first all-female aircrew.

GENERAL VAUGHT & THE WOMEN'S MEMORIAL

Prior to 1967, it was against the law for a woman to become an Air Force general, but Wilma Vaught earned her star and went on to establish the Women In Military Service For America Memorial at the gateway to Arlington National Cemetery.[115]

Women have served during every one of our wars, a fact that surprises most Americans. When her husband was killed in battle, Margaret "Captain Molly" Corbin rushed to replace him at the cannon defending Ft. Washington during the Revolutionary War. From the Civil War through Vietnam, thousands of women served in the armed forces, often as nurses.[116]

Following the military's success with racial desegregation, opportunities also began to increase for women in uniform. The service academies admitted females beginning in 1976. In 1999, Eileen Collins commanded a Space Shuttle mission. Former CAP cadet Kim Campbell earned the Distinguished Flying Cross for heroism as an A-10 pilot in Iraq in 2003. Another former cadet, Nicole Malachowski, became the first woman to fly as an Air Force Thunderbird, in 2006.[117]

"When I graduated from college in the 1950s, women were supposed to be teachers, nurses, get married, be secretaries," remembers General Vaught. "What I wanted to do was be in charge, but I would never have a chance to move into management [in private industry]."[118] While other organizations refused to recognize her abilities, the Air Force granted Vaught an opportunity to lead. She was one of the first female officers to earn flag rank, and General Vaught's record of achievement paved the way for other women to succeed.

The Women In Military Service For America Memorial honors all women, past, present, and future, who serve. In dedicating the $22 million memorial, General Vaught said, "They are so proud and yet have not been recognized... We had to tell their story because it had to be told – it never had been told before."[119]

Brig Gen Wilma Vaught USAF (Ret.) visiting with CAP cadets

AMERICA'S INCREASING DIVERSITY

OBJECTIVES:

42. Defend the claim that America's diversity is increasing.
43. Explain how America's increasing diversity will affect leaders.

Respect for diversity will become even more important in the future. Demographics, the statistical study of people, shows that the United States is expected to become an older and more racially and culturally diverse population. Individuals who are uncomfortable in diverse environments today will need to become better skilled in working in a diverse culture. This sampling of data (see right) illustrates the point:[120]

	2005	2050	Difference
Whites (non Hispanic)	67%	47%	- 30%
Hispanics	14%	29%	+ 107%
Blacks	13%	13%	unchanged
Foreign Born	12%	19%	+ 58%
Workers (Aged 18-64)	63%	58%	- 8%
Seniors (Aged 65+)	13%	20%	+ 53%
Total US Population	310 mil	440 mil	+ 42%
Total World Population	6.5 bil	8.9 bil	+ 37%

PREJUDICE, HATRED, AND THE LEADER

OBJECTIVES:

44. Define the term "prejudice."
45. Define the term "harassment."
46. Define the term "retaliation."
47. Describe the role a leader has in fighting prejudice and hatred.

What is prejudice? The answer lies in the word itself. ***To be prejudiced means to pre-judge someone.*** Making assumptions about an individual just because they are male or female, black or white, Eskimo or Swahili, practice a certain religion, or display certain personal traits is a form of prejudice. Although assuming good thoughts about someone is a sign of prejudice – your older sister is athletic, I bet you're athletic, too – when people speak of prejudice they are usually referring to a form of hatred and distrust.

Prejudice is important because hateful feelings too often give rise to hateful actions, like harassment. ***Harassment is unwelcome conduct.***[121] It's the attitude that says, "I'll make life difficult for someone." In the workplace, or in a volunteer group like CAP, the organization can be made responsible for harassment, especially if it comes from a supervisor, like an NCO or officer.[122] To clarify, petty slights and annoyances are usually not examples of illegal harassment, although most people consider them to be socially unacceptable.

Because America is built upon democratic values, society not only opposes prejudice and harassment, it opposes retaliation. ***Retaliation is when someone seeks revenge against someone who objects to harassment or discrimination.***[123] When a boss or employer tries to fire, demote, or deny an award to someone who speaks out against discrimination, that boss or employer is guilty of retaliation.

Leaders are expected not to display signs of prejudice or harass or discriminate against other people. Further, ***because leaders are responsible for their teams, leaders are expected to create an atmosphere that welcomes everyone.***[124] Supervisors who remain silent in the face of harassment and hatred can be held responsible for their failure to lead.

> **"Hateful feelings too often give rise to hateful actions."**

"First they came for the Jews, and I did not speak out because I was not a Jew. Then they came for the Communists, and I did not speak out because I was not a Communist. Then they came for the trade unionists, and I did not speak out because I was not a trade unionist. <u>Then they came for me, and there was no one left to speak out for me.</u>"

Rev. Martin Niemöller
Survivor of the Nazi concentration camp
known as Dachau[125]

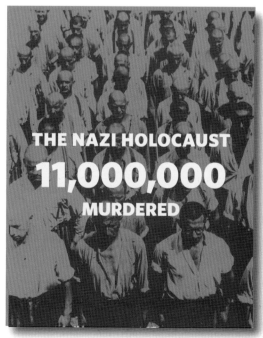

THE NAZI HOLOCAUST
11,000,000
MURDERED

FIVE WAYS TO FIGHT HATE

OBJECTIVE:

48. Describe the five-step process for fighting hate.

Good leaders see the wrong in prejudice, but how can they effectively counter hatred and bigotry? This five-step process can help leaders show their support for democratic traditions and CAP's Core Value of Respect:[126]

1.   Rise Up. *Inaction in the face of prejudice is almost as bad as the hatred itself.* When a leader fails to act, he or she signals, perhaps unintentionally, support for bigotry. Leaders need to not let hate go unchallenged.

2.   Pull Together. Most people have been raised to reject hatred and bigotry. Leaders who see hatred can expect that at least some other members of the team, if not all, will oppose harassment and discrimination. *Sometimes it takes just one brave individual to rally others who have remained quiet in the face of hatred.*

3.   Speak Out. *Leaders need the courage to act.* They can personally challenge the individuals who harass others. Sending the message that "we don't condone that behavior," or "those aren't our values" is important. Also, speaking out means telling other leaders in positions of authority about the problem.

4.   Support the Victims. *People who experience harassment and other types of hatred may need support.* Depending on the situation, that support can range from counseling and health or legal services to simply having someone they can talk to, following the wingman principle discussed in chapter two. As leaders try to help, it is important they not attempt to speak for the victim or allow their good intentions to re-victimize a victim.

5.   Teach Tolerance. If there is a silver lining to prejudice and bigotry it is that *tough times give leaders an opportunity to teach tolerance.* School and the CAP Cadet Program are environments where young people can learn from their own and one another's mistakes. Leaders ought not miss the chance to eradicate ignorance.

DRILL & CEREMONIES

As part of your study of this chapter, you will be tested on your ability to lead a squadron in drill and ceremonies. Ask an experienced cadet to help you learn the procedures governing the four areas of squadron drill listed below. For details, see the *Drill and Ceremonies Manual* available at capmembers.com/drill.

From the Air Force Drill & Ceremonies Manual, Chapter 5

Forming the squadron in line
Aligning the squadron
Inspecting the squadron
Dismissing the squadron

ENDNOTES

1. Richard Halverson, "Personality," Flathead Valley Community College, http://home2.fvcc.edu/~rhalvers/psych/Personality, retrieved Nov 2009.

2. Steven Pinker, "Why Nature & Nurture Won't Go Away," *Daedalus*, Fall 2004, 10.

3. Ibid, 3.

4. Ibid, 5.

5. John Locke, "Essay Concerning Human Understanding," in *A History of Western Philsophy*, Bertrand Russell, ed., (New York: Touchstone, 1972 ed.), 722.

6. Attributed to Jose Ortega Gasset by Pinker, 1.

7. John B. Watson, *Behaviorism*, (Chicago: University of Chicago Press, 1924), 104.

8. Pinker, 3.

9. Attributed to Stephen Jay Gould by Pinker, 4.

10. *Gattaca*, directed by Andrew Niccol, Columbia Pictures, 1997.

11. *Rudy*, directed by David Anspaugh, TriStar Pictures, 1993.

12. Rudy Ruettiger, http://www.rudyinternational.com/inspirational.cfm, retrieved Nov 2009.

13. Child Development Institute, "Birth Order," http://www.childdevelopmentinfo.com/development/birth_order.htm, retrieved Oct 2009.

14. Jeffrey Kluger, "The Power of Birth Order," *Time*, Oct 17, 2007.

15. Ibid.

16. Ibid.

17. Frank J. Sulloway, "Birth Order & Intelligence," *Science*, 22 Jun 2007, 1712.

18. Ibid, 1711.

19. John Stossel, "Does Birth Order Determine Personality?," *ABC News*, May 21, 2004.

20. Ibid.

21. Child Development Institute.

22. Attributed to Marianne Williamson.

23. R. J. House, "A 1976 Theory of Charismatic Leadership," in *Leadership: The Cutting Edge*, J.G. Hunt, ed., (Carbondale, IL: So. Illinois Press, 1977), 189-204.

24. Afsaneh Nahavandi, *The Art & Science of Leadership*, (Upper Saddle River, NJ: Prentice Hall, 2003), 224.

25. Ibid, 233.

26. Lynn Offerman, "When Followers Become Toxic," *Harvard Business Review on the Mind of the Leader*, (Boston: Harvard, 2005), 39.

27. Attributed to Peter Drucker.

28. Attributed to Warren Bennis.

29. Donald T. Phillips, *The Clinton Charisma*, (New York: Macmillan, 2008), 229.

30. Richard Berke & Don Van Natta Jr., "...President Told His Closest Aides the Painful Truth," *New York Times*, Aug 18, 1998.

31. "Jonestown: The Life & Death of the Peoples Temple," *American Experience*, PBS, 2006.

32. James MacGregor Burns, *Transforming Leadership*, (New York: Grover Press, 2003), 26.

33. *Forrest Gump*, directed by Robert Zemeckis, Paramount Pictures, 1994.

34. Duen Hsi Yen, "Johari Window," http://www.noogenesis.com/game_theory/johari/johari_window.html, retrieved Nov 2009.

35. Paul Hersey, Kenneth H. Blanchard, & Dewey E. Johnson, *Management of Organizational Behavior*, 7th ed., (Prentice Hall: Upper Saddle River, NJ), 304.

36. "Interactive Johari Window," http://www.kevan.org/johari, retrieved Nov 2009.

37. MindTools, "The Johari Window," http://www.mindtools.com/CommSkll/JohariWindow.htm, retrieved Nov 2009.

38. Ibid.

39. Hersey, Blanchard, 307.

40. Air Force JROTC, *Leadership Education II*, (Boston: McGraw Hill, 2005), 149.

41. Paul Tieger & Barbara Barron, *Do What You Are*, 4th ed., (New York: Little, Brown & Co., 2007), 14-25.

42. Attributed.

43. Attributed.

44. *Star Trek*, created by Gene Roddenberry, 1966, http://www.imdb.com/title/tt0060028/quotes, retrieved Nov 2009.

45. Otto Kroeger & Janet Thuesen, "The Typewatching Profiles," excerpted from *Type Talk*, (New York: Bantam, 1989).

46. Air Force JROTC, 151.

47. Morton Hunt, *The Story of Psychology*, (New York: Anchor, 2007), 581.

48. Abraham Maslow, "A Theory of Human Motivation," in *Classics of Organization Theory*, 4th ed., Jay Shafritz and J. Steven Ott, eds., (New York: Harcourt Brace, 1996), 164-168.

49. Tom Wolfe, *The Right Stuff*, (New York: FSG, 1983), 244.

50. Robert Browning, "Fra Lippo Lippi," line 54.

51. Attributed by the BBC.

52. Maslow, 168.

53. "Chapel of the Pieta," http://saintpetersbasilica.org/Altars/Pieta/Pieta.htm, retrieved Nov 2009.

54. Hal Rainey, *Understanding & Managing Public Organizations*, 3rd ed., (San Francisco: Jossey Bass, 2003), 32.

55. Ibid, 32.

56. Ibid, 32.

57. "Pavlov's Dog," Nobelprize.org, http://nobelprize.org/educational_games/medicine/pavlov/.html, retrieved Nov 2009.

58. Jane Weightman, *Managing People*, 2nd ed., (London: Chartered Inst. of Personnel & Development, 2004), 35.

59. Ibid, 35.

60. "Types of Operant Conditioning," ChangingMinds.org, http://changingminds.org/explanations/behaviors/conditioning/types_conditioning.htm, retrieved Nov 2009.

61. Stanley Milgram, *Obedience to Authority*, (New York: Taylor & Francis, 1974).

62. Ibid, 21.

63. Ibid, 20.

64. Ibid, 6.

65. Attributed to Albert Einstein.

66. *Uniform Code of Military Justice*, 10 U.S.C., Chapter 47.

67. Milgram, 8.

68. Attributed to Max Lucado.

69. Louis Imundo, *The Effective Supervisor's Handbook*, 2nd ed., (New York: American Management Association, 1991), 182.

70. Phebe Cramer, *Protecting the Self...*, (New York: Guilford Press, 2006), 7.

71. Air Force JROTC, 116.

72. Cramer, 6.

73. Richard Niolon, Ph.D., "Defenses," Dec 1999, http://www.psychpage.com/learning/library/counseling/defenses.html., retrieved Nov 2009.

74. Cramer, 4.

75. Ibid, 5.

76. Niolon.

77. Cramer, 7.

78. George E. Vaillant quoting the DSM-III in *Ego Mechanisms of Defense*, (New York: American Psychiatric Pub., 1992), 237.

79. Ibid, 238.

80. Ibid, 238.

81. Niolon.

82. Vaillant, 238.

83. Ibid, 237.

84. Ibid, 238.

85. Niolon.

86. Ibid.

87. Robert Heller, *Dealing With People*, (New York: DK Publishing, 1999), 46.

88. Harry Webne-Behrman, "Conflict Resolution," University of Wisconsin-Madison, http://www.ohrd.wisc.edu/onlinetraining/resolution/index.asp, retrieved Nov 2009.

89. James A. Autry, *The Servant Leader*, (New York: Three Rivers, 2004), 189.

90. Heller, 7.

91. Autry, 187.

92. Ibid, 172.

93. Ibid, 191.

94. Nobel Peace Laureate Project, "Jimmy Carter," http://www.nobelpeacelaureates.org/pdf/Jimmy_Carter.pdf, retrieved Nov 2009.

95. Central Intelligence Agency, "North Korea," *CIA World Factbook*, (Washington: CIA, 2006), 304.

96. Ibid, "Germany," 215.

97. Thom Shanker, "Commanders Are Faulted on Assualts at Academy," *New York Times*, Dec 8, 2004.

98. Daniel Pulliam, "Pentagon Blames AFA Leaders for Sexual Misconduct Scandal," Government Executive, Dec 8, 2004.

99. Autry, 193.

100. Webne-Behrman.

101. Civil Air Patrol, *Leadership for the 21st Century*, vol. 2a, (Maxwell AFB, AL: CAP, 1993), 25.

102. Ibid, 25.

103. Webne-Behrman.

104. Karen Breslau & Katrina Heron, "Bill Clinton," *Wired*, Dec 2000.

105. Ibid.

106. businessdictionary.com

107. ibid.

108. Tolerance.org, "Declaration of Tolerance," Southern Poverty Law Center, http://www.tolerance.org/about, retrieved Nov 2009.

109. Ibid.

110. President Harry Truman, *Executive Order 9981*, July 26, 1948.

111. LCDR Todd Varvel USN, "Ensuring Diversity is Not Just Another Buzz Word," (Maxwell AFB, AL: Air Command & Staff College, 2000), 5.

112. CMSAF Rodney McKinley, "Air Force Diversity," *The Enlisted Perspective*, Apr 27, 2009.

113. Robert Neprud, *Flying Minute Men*, (New York: Duell, Sloan & Pearce, 1948), 198-199.

114. Ibid.

115. Women in Military Service for America Memorial, http://www.womensmemorial.org/, retrieved Nov 2009.

116. Ibid.

117. National Headquarters, Civil Air Patrol.

118. Michel Martin, "Wisdom Watch: Wilma Vaught," *National Public Radio*, July 18, 2007.

119. *New York Times*, "Memorial for Women in the Military is Dedicated," Oct 19, 1997.

120. Jeffrey Passel & D'Vera Cohn, "U.S. Population Projections: 2005-2050," (Washington: Pew Research Center, 2008).

121. U.S. Equal Employment Opportunity Commission, "Discriminatory Practices," http://www.eeoc.gov/abouteeo/overview_practices.html, retrieved Nov 2009.

122. Ibid.

123. Ibid.

124. Ibid.

125. Attributed in Willoughby (see note 123).

126. Brian Willoughby, "10 Ways to Fight Hate on Campus," (Montgomery, AL: Southern Poverty Law Center, 2005).

CHAPTER 7
LEADERSHIP SCHOOLS OF THOUGHT

"BEGIN WITH THE END IN MIND. Begin today with the image, picture, or paradigm of the end of your life as your reference or [standard] by which everything else is examined. Each part of your life – today's behavior, tomorrow's behavior, next week's behavior, next month's behavior – can be examined in the context of the whole, of what really matters to you. By keeping that end clearly in mind, you can make certain that whatever you do on any particular day does not violate the criteria you have defined as supremely important, and that each day of your life contributes in a meaningful way to the vision you have of your life as a whole."[1]

CHAPTER GOALS

1. Appreciate the benefit of raising your emotional intelligence.

2. Understand how you can transform your team.

3. Appreciate different approaches to leadership.

CHAPTER OUTLINE
In this chapter you will learn about:

EMOTIONAL INTELLIGENCE
  Self-Awareness
  Managing Emotions
  Self-Motivation
  Empathy for Others
  Interpersonal Skills

TRANSFORMATIONAL & TRANSACTIONAL LEADERSHIP
  Idealized Influence
  Inspirational Motivation
  Intellectual Stimulation
  Individualized Consideration
  Contingent Reward
  Management by Exception
  Laissez-Faire

POWER
  Definitions of Power
  Power Within Organizations

BUILDING A LEARNING ORGANIZATION
  Systems Thinking
  Personal Mastery
  Shared Vision
  Team Learning
  Mental Models

LEADERSHIP STYLES
  Situational Leadership Theory
  Path-Goal Model
  The Leadership Grid

Even in a large corporation, success starts with a single person – you. An organization is only as strong as the personnel with which it is made. *People make the organization.*

You affect the organization through your attitude, spirit, and ability to communicate your thoughts and feelings.

Once you can manage yourself, you are ready to manage a small group, such as a flight of cadets.

How can you make your group successful? What type of leader will you be?

Your actions and decisions will contribute to the success or failure of your group. Thinking collectively, if all the groups in an organization fail, the company will suffer and could even crumble completely.

So you see how important you are. And you see how important your group's success is. *The future of your organization does rely very much on you and how you lead your group.*

EMOTIONAL INTELLIGENCE

OBJECTIVES:
1. Define "emotional intelligence."
2. Name the five components of emotional intelligence.

If you don't really understand yourself, how can you expect to understand your followers and to lead others? This was one of the main points of chapter 6. Let's consider it further.

Much of our personalities are conveyed to others through emotions. If we are generally calm and easy-going, others recognize that in us. At the same time, if we all of a sudden burst out in anger or lose our temper easily, others notice these emotions as well.

What's Your Emotional Intelligence?
Emotional intelligence refers to how well you handle yourself and your relationships as a leader. What can you tell from this cadet's appearance to indicate how he feels?

And yet the issue is not only with our anger or loss of temper.

What if we don't realize how we appear when we show our frustration and anger? What if we don't even realize that we are frustrated and angry?

If we don't understand how we feel, then, when our anger hurts the feelings of a peer or follower, we are absolutely oblivious to the fact that we've wronged somebody.

And if we're suffering from anxiety or sadness and don't recognize it, we may continue to harm ourselves with negative feelings.

Touchy-Feely?
The name is misleading. Emotional intelligence sounds touchy-feely, like leadership's equivalent to the Teddy bear. It's not. Rather, it's about having respect for how emotions affect leadership. After all, having vision, motivating people, and helping them see the essence of your ideas involves the emotions. Not surprisingly, emotional intelligence is studied at the Air Force's highest school, Air War College.

If we want to lead a stable team, we must be stable ourselves and recognize our own emotions, how and when we show emotion, and emotion in others as well.

Solidifying your emotional intelligence (EI) is the first step to leading others. As one expert puts it, ***emotional intelligence is the "intelligent use of emotions: you intentionally make your emotions work for you by using them to help guide your behavior and thinking in ways that will enhance your results."***[2]

Other researchers also believe that emotional intelligence is a key ingredient in leadership, especially in transforming others.

EI (emotional intelligence) might be a central factor in several leadership processes, particularly in the development of charismatic and transformational leadership, where the emotional bond between leaders and followers is imperative. Being able to empathize with followers can further allow a leader to develop followers and create a consensus.[3]

Another expert asserts that individuals with high emotional intelligence can remain motivated even when facing challenges. They can detect emotion in others, feel empathy for those who may be hurting, and put themselves in the place of team members who may need counsel.[4]

Experts in the field identify five primary aspects of emotional intelligence:

(1) self-awareness
(2) managing emotions
(3) self-motivation
(4) empathy for others
(5) interpersonal skills.[5]

As you look at each item, consider whether or not you are already practicing the aspect of emotional intelligence or if it is an area in which you can improve.

INCREASING SELF-AWARENESS

OBJECTIVES:
3. Define "self-awareness."
4. Define "appraisal."
5. Define "self-fulfilling prophecy."

Knowing yourself is important. Are you generally optimistic? Do you tend to see the good in others rather than their flaws? What is your personality like? Are you shy or outgoing? Are you generally happy and content? Or are you often sullen and depressed? Do you ever get angry? Do you raise your voice to a high volume when you are upset? Perhaps you are calm and even the biggest shocks barely cause your heart rate to rise.

Whatever the case may be, ***if you understand your emotions, you can also begin to change them*** if you realize that you are stressed. You can begin to understand how the way you feel can affect those around you.

Self-awareness, according to one expert, means to be "aware of both our mood and our thoughts about that mood."[6] When you recognize your moods, you can more readily alter them, according to experts. In other words, if you can realize you are not comfortable with a situation, you can begin to decide what remedy can change or erase your difficulty.

> ### WHAT DOES THIS MEAN FOR ME?
> #### *Applying Emotional Intelligence*
>
> **EI can be a useful tool for self-development. Here are some areas to work on:**
>
> ★ Learn about your strengths and weaknesses by taking self-assessments, engaging in honest reflection, and seeking and listening to feedback.
>
> ★ Work on controlling your temper and your moods; stay composed, positive and tactful when facing difficult situations.
>
> ★ Integrity is a choice; stay true to your word and commitments.
>
> ★ Set challenging goals and be willing to work hard to achieve them; admit your mistakes and learn from them.
>
> ★ Build relationships with others and develop your network.
>
> ★ Pay attention to those around you; be concerned about their well-being and feelings.

COMPONENTS OF EMOTIONAL INTELLIGENCE

Component	Description
Self-awareness	Being aware and in touch with your own feelings and emotions
Managing emotions	Being able to manage various emotions and moods by denying or suppressing them
Self-motivation	Being able to remain positive and optimistic
Empathy for others	Being able to read others' emotions accurately and putting yourself in their place
Interpersonal skills	Having the skills to build and maintain positive relationships with others

This type of recognition by self-aware people tends to make them more independent, in better mental and emotional health and more positive in their view of life.[7] When a self-aware person gets into a bad mood, he can get out of it and move on with life sooner.[8]

Appraisals. Part of knowing yourself involves the way in which you characterize, or describe, how you feel about events that you face throughout a day. Another word for this is *appraisals – "all the different impressions, interpretations, evaluations, and expectations you have about yourself, other people, and situations."*[9]

Do you think you've never made an appraisal? No? Think again.

It's 3 p.m. You are at school or home-schooling in your house. All should be well except you have to give a speech at your squadron meeting this evening. Oh no! You are so nervous. You hate to speak in front of others, and in fact you just transferred to this squadron, so you'll have a room full of unknown faces awaiting your every word. Yikes! How are you feeling? You think, "I am going to flop. I'm going to give the worst speech ever. I just can't do this."

Self-Fulfilling Prophecies. Your appraisal or feeling is that you will perform terribly during your speech. And when you fill yourself with additional anxiety and stress, often you increase the chance of actually presenting a terrible speech. It's called a self-fulfilling prophecy.[10] *A self-fulfilling prophecy occurs when your prediction or expectation leads to your fears or hopes becoming real.*[11]

The good news is that your appraisals don't have to be negative or frightful. You can replace your fearful thoughts about

Believe in Yourself
Confidence goes a long way. If you anticipate your speech being successful, it likely will be.

your speech with positive thoughts. "I'll give a great speech; I might be a little nervous, but I know I've researched this topic well, and it will be exciting to share what I know with my new squadron peers." By encouraging yourself in such a way, you will be able to relax more and be more comfortable in the moments leading up to your speech and throughout your delivery.

Merely thinking positively helps put you at ease and rids you of unnecessary anxiety and fear. It's called the power of positive thinking!

MANAGING YOUR EMOTIONS

OBJECTIVES:
6.   Define "managing emotions."
7.   Define "automatic thoughts."
8.   Explain what "constructive inner dialogue" means.

Emotional intelligence is complex, but understanding it will benefit you throughout your life. Self-awareness is important, and knowing yourself goes hand-in-hand with managing your emotions. Managing emotions isn't just about anger management.

Emotional Balance
These cadets' smiles reflect a healthy competition between the two.

There are times when one is filled with anger and when keeping this emotion in control becomes important to your health and the well-being of those around you. Of course, many of us also have been in situations in which solemnity was required, but we resisted the temptation to laugh, smile and show happiness. This might have occurred at a funeral in which the mood was solemn, but you suddenly thought of something that made you want to chuckle. Being able to control your anger and even your glee is critical.

Solemnity.
the quality of being serious and dignified

Some of us might have met people who, if upset, can remain angry for a day, even a week, holding a grudge and stewing over the insult they received. But think of what being this angry does to one's concentration. If your mind is filled with angry thoughts for a day or a week, it will be much more difficult to focus on the test you have tomorrow, on the chores you might have at home, or on enjoying the good things that may happen to you. ***Not managing emotions can lead even the nicest people into dangerous and unhealthy situations.***

Inner Dialogue. *One way to manage emotions is through inner dialogue, or talking to yourself.*[12] Don't worry that people will think you are crazy if you talk to yourself; in any case, these can be silent thoughts within your mind that help you respond to emotion.

Take an instance when you suddenly become angry at a friend. This person you had always trusted suddenly disappoints you by telling a secret to others. You are so upset that you feel like ending your long friendship immediately. But no one's perfect, and this might be the perfect moment to escape to a quiet place for some inner dialogue. You might say to yourself, "I am so angry, but being this upset really makes me sad and depressed. I don't want to feel this way, and there was the time when I made a mistake and upset him/her, too. Maybe I can just go get a milkshake to help me feel better, then tell my friend that I was really hurt by the breach of trust."

Stopping to think and reflect helped you think of a short-term solution to improve your mood (getting a snack), aided you in remembering the time where you erred, and allowed you to decide on a mature course of action (letting your friend know how you feel). On the other hand, *when you don't pause to consider your emotional state of being, your emotions can spin out of control.*

Automatic Thoughts. *Your emotional tailspin often might begin with your automatic thoughts, "thoughts that spontaneously pop out."*[13] We've all thought automatically. For instance, your mother might be driving and another motorist cuts in front of her car. Her first reaction might be: "Oooohhhh. I wish he'd have his license taken away forever." Or every time you are at a SAREX (search-and-rescue exercise), the same peer boasts that he is the best ELT finder. One might think, "The nerve of this guy. He is arrogant. I really dislike this person."

After a while, these *negative thoughts can overpower you and dominate your thinking.* You can convince yourself that your thoughts are accurate and valid when, in fact, your judgments of another person may be very untrue.

To be a good leader, you must be aware of your emotions and manage them to avoid being a knee-jerk leader flying with your brain on "autopilot."

YOUR BRAIN ON AUTOPILOT

Automatic thoughts are like autopilot. It's as if something else is in control of your brain, not the real you.

Situation: You're working at a convenience store to earn extra money. A customer speaks to you like you are a five-year-old. Your first reaction is to curse at him.

Worst plan: You do curse at him. Your boss finds out and fires you on the spot. You had cursed at his best customer, who vowed never to visit the store again.

Better plan: Don't be oversensitive. You know you are mature. Resist cursing and ask the valued customer how his day is going. He'll see you are older than you look. You will gain respect and keep your job.

MOTIVATING YOURSELF

OBJECTIVE:

9.   Define "self-motivation."

Motivating yourself can begin before disaster or other setbacks strike. You've seen the importance of being self-aware and of managing emotions. Often your self-awareness and ability to manage emotions come into play when things happen to you – you lose a game, a close friend dies, someone offends you. At these points your emotional management helps dictate how you will react to these occurrences.

Your emotional intelligence is closely related to your level of motivation, or your willingness to complete tasks. For instance, if you feel so doomed by the size of a task – whether it be writing an essay or mowing a lawn – perhaps you will not even be able to start. You will set yourself up for failure before you even start, which will decrease your ability to feel good about yourself.

You can motivate yourself through positive thinking by using motivational statements, using mental imagery, and setting meaningful goals. In other words, **good leaders are proactive**. That is, they take initiative to change their world rather than wait for change to happen to them.

> "Negative thoughts can overpower you and dominate your thinking."

MOTIVATIONAL STATEMENTS

OBJECTIVE:

10.   Construct a self-statement.

Motivational self-statements are simple expressions of belief in oneself. A motivational statment will fortify your optimism, tenacity and resiliency.[14]

Earlier, we spoke of inner dialogue, or talking to yourself. Motivational statements are similar. Often we lack motivation because we think a task is too overwhelming. We don't believe we can do it, and we get stuck and never start because we feel overpowered. We tell ourselves, "I can't do it. This is too much." Thoughts like this confirm our initial feelings of inadequacy. Rather, we can use inner dialogue in the opposite direction, and say to ourselves, "I can do this. I can succeed. This

The Little Engine That Could

Watty Piper's famous children's story, *The Little Engine That Could*, offers an excellent example of a motivational statement. "I think I can, I think I can," cried the little engine. That attitude is what propelled the train to the top.

project isn't too big for me." This is the starting block. If we can get started through positive motivation, often half the battle is won.

MENTAL IMAGERY

OBJECTIVE:
11.   Explain what "mental imagery" is.

Remember, when you understand how to motivate yourself, you gain a better grasp of how to regulate, or control, your emotions.

You take charge of your emotions, rather than them controlling you.

Through motivation, you can change your attitude from one that is uninspired to one that is thrilled and ready to tackle the world.

And motivation is such a powerful tool. Not only can you use positive inner dialogue to motivate yourself (using words and language,) but you can also use pictures or images that are positive as well.

You can use mental imagery to picture yourself in a situation.[15] In other words, if you have a presentation to deliver, you could actually, within your mind, see yourself walking to the front of a room of your peers, stepping behind the podium, making a joke to relax yourself, then confidently delivering an excellent speech. You will gain confidence and become more motivated when you envision yourself as a success.

In addition, you also can use mental imagery to observe and mentally record a peer performing a drill movement with skill, then try to imitate your peer's actions.[16] The mind is an excellent ally for staying motivated through whatever life challenges you with.

Make a mental picture
You want to be named Cadet of the Year. To help yourself reach your goal, you might use mental imagery. You picture yourself on the stage at CAP's annual convention, receiving a huge trophy from the National Commander.

"You will gain confidence and become more motivated when you envision yourself as a success."

RIGHT-SIZED GOALS

OBJECTIVE:
12.   Identify two errors one can make when setting goals.

Your motivation also is decreased or increased depending on how you set goals. ***If you set goals that can actually be met, you will have more inspiration to work hard and steadily.***

There are two errors you can make when setting goals: You can set them too high, or you can set them too low.[17]

You want to be an extremely well-performing cadet. Perhaps your first idea is to envision being the very best cadet out of all of Civil Air Patrol's more than 20,000-strong cadet corps. That's an overwhelming goal. It's not impossible to reach, but for a beginning cadet, it's pretty lofty. But being the top cadet in your home squadron is a very attainable goal. The idea is to give yourself a goal that thrills you, encourages you to keep striving, and one that you can envision accomplishing.[18] ***Goals that encourage you to proceed confidently but do not overwhelm you are the best.***

Flow as Focus
Are they so absorbed in the competition that they've forgotten about everything else? Olympic Gold Medalists Apolo Ohno (top) and Shaun White (bottom) demonstrate incredible focus while making their tremendous athleticism look easy. They exemplify what leadership experts call "flow."

GO WITH THE FLOW

The ultimate goal of self-motivation, according to one expert, is to achieve flow. ***Flow is the feeling people enjoy when they are so absorbed in a task that they forget about all other worries.*** "Flow is emotional intelligence at its best. ... flow represents perhaps the ultimate in harnessing the emotions in service of performance and learning."[19]

These Olympians didn't wake up one day and decide to try out for the Olympics. They set incremental goals for themselves, practiced until they perfected each step, and eventually won the gold. While it may look like they were born flying through the air or speeding on ice, in reality they've achieved flow.

Empathy in Action: The Candy Bomber

An Air Force officer having a reputation for tremendous empathy was 1st Lt (later Col) Gail Halvorsen.

The Soviets had blockaded Berlin shortly after WWII. In response, the U.S. and its allies flew thousands of tons of food to that captive city.

But Halvorsen was not content to simply fly his mission and call it a day. He was a man of empathy. He felt for the children of Berlin who had suffered terribly during WWII and were continuing to suffer under Soviet rule.

Halvorsen collected as much candy as he could find, fashioned small handkerchiefs into parachutes, and dropped treats to gleeful kids below. Soon other pilots followed his lead, with the help of America's candy manufacturers.

The Candy Bomber's empathy generated incalculable feelings of goodwill and showed a generation of Berliners that the U.S. pilots were the good guys. He is still affectionately known to many as *Onkel Wackelflügel* or "Uncle Wiggly-Wings."

EMPATHY FOR OTHERS

OBJECTIVE:

13.   Define "empathy."

Empathy is understanding, being aware of, and being sensitive to the feelings, thoughts, and experiences of another. The old saying, "you don't know someone until you walk a mile in his shoes" represents empathy well. Because emotions can be contagious, leaders need empathy. Leaders need to recognize how others must feel.[20]

Have you even been around someone who is upset? Maybe you were feeling fine until a conversation with him or her? Then your teammate's anger makes you mad, too. When you begin imitating others' behaviors – yelling back at someone who is yelling at you – you can start to feel agitated as well.[21]

Empathy's Potential to Help. If leaders are to help their people, leaders need empathy. You can learn to detect emotions in others. For example, instead of sitting idly by while a team member is hurting, you can try to help.[22] Say your teammate is speaking slowly, in a quiet tone, and sitting in his chair with his head down – he's not acting enthusiastic like he normally does, so you know something is wrong and that his emotions can affect the entire team. By recognizing that he may be tired or even depressed, you are able to intervene. You can then ask if he would feel better to share his disappointment or frustration, or you could refer him to a counselor. Either way, **by learning to see emotions in a team member, you support the entire team.**

Understanding emotions in others can really set you apart as a leader because you can help followers through tough times. Fortunately, followers won't always need such support, but when they do, you can be their rock.

Empathy & The Discipline Sandwich. As another example, consider having to counsel a cadet regarding an infraction he committed during an activity. Suppose the cadet is a new cadet who has already shown himself to lack confidence. You know he will be nervous to take part in a counseling session. Recalling what you learned about the "discipline sandwich," (page 21), you begin the session by making positive remarks about him so that he won't be so nervous and can exit the session prepared to put his error behind him and move on. In short, **by predicting another person's emotions, you can direct outcomes in a positive direction.**[23]

> "When your team needs help, you can be their rock."

INTERPERSONAL SKILLS

OBJECTIVE:

14. Describe what is meant by the term "interpersonal skill."

Interpersonal skill involves developing and maintaining positive relationships with others. Leaders who possess excellent interpersonal skills are said to be good at networking. That is, they are good at meeting new people, relating to them, and finding ways to build mutually beneficial relationships. Empathy provides a foundation for interpersonal skills because being able to recognize how others feel is a prerequisite for developing a good relationship with them.

If a leader possesses strong interpersonal skills, he or she is said to have high **interpersonal intelligence**, which one

Networking Opportunity
Here four cadets meet with Vice President Joseph Biden (then a Senator). Perhaps they told Mr. Biden of their desire to earn service academy nominations, explored internship opportunities on Capitol Hill, or reminded him of how CAP serves America. Regardless, cadets who develop their interpersonal skills become successful at networking.

The Leader as Negotiator
Negotiation is an important aspect of interpersonal skill and it's harder than it looks. On one hand, the negotiator must be steadfast on matters of high principle. On the other hand, he or she must be flexible and open-minded. The U.S. Secretary of State (here, Hillary Clinton, right), is the chief diplomat and negotiator for U.S. foreign policy.

expert defines as "the ability to understand other people: what motivates them, how they work, how to work cooperatively with them."[24] What follows is a look into the four components of of interpersonal intelligence:[25]

Organizing Groups. Vital for leading others, this skill refers to the ability to put a group together and inspire collective action. Military leaders and organization directors demonstrate this talent.

Negotiating Solutions. People who have skill in negotiation are very successful in making deals and settling disputes. In children, one might see a negotiator on the playing field. He or she is the one trying to stop a fight or settle a disagreement. In politics, every president of modern times has struggled to negotiate a peace in the Middle East.

Personal Connection. Our emotional intelligence expert calls this the art of relationship. It involves empathy and connecting. The ability to connect with others develops out of showing concern for them, getting to know their hopes and fears, and understanding their unique backgrounds. People with this skill can read emotion in others not just from what is said but also from the facial expressions of others.

Social Analysis. A curiosity about and an ability to notice people's feelings and concerns are the mark of social analysis. If you can detect the feelings of others, you can help them through a problem. Therapists and counselors are especially strong in this ability.

All of the above skills point back at the beginning of our discussion of emotional intelligence – it starts with you. If you are aware of emotion in yourself and if you can control your emotions, you can be a rock of stability for others.

Leader has a high degree of influence upon follower	*Leader has a low degree of influence upon follower*	*Leader has little or no influence upon follower*
Transformational Leadership	**Transactional Leadership**	**Laissez-Faire Leadership**
★ Idealized Influence	★ Contingent Reward	★ Non-leadership
★ Inspirational motivation	EXAMPLE: Win a $10 bonus for every cadet recruited	★ Little if any leader-follower interaction
★ Intellectual Stimulation		EXAMPLE: Commander remains in office, never interacting with the team, apparently content to allow troops to do as they please
★ Individualized Consideration	★ Management by Exception	
EXAMPLE: Focus on people, values, power of persuasion, winning of "hearts and minds"	EXAMPLE: receive demerits and reprimand for not meeting quota	

TRANSFORMATIONAL & TRANSACTIONAL LEADERSHIP

OBJECTIVES:

15. Define "transformational leadership."
16. Define "transactional leadership."
17. Explain the difference between transformational and transactional leadership.

The best leaders are ones who focus on getting the team to profess the right values. In so doing, the leader "transforms" the team into a force for good, ready to work hard to make those values come to life.[26]

When Ronald Reagan called the Soviet Union the "evil empire," he was telling the world something about his values. Partly because of that speech, people from around the world came to see the Soviet Union as a nation that stood in the way of the common good. They followed Reagan's lead and joined him in pressuring the Soviets to change their ways. Reagan had transformed people's values.

The study of transformational leadership goes hand-in-hand with transactional leadership. Let's take a closer look at each.

Transformational Leadership. ***In transformational leadership, a person strives to heighten the motivation and morality of himself and his followers.***[27] This leader works to help followers "reach their fullest potential."[28] One expert's definition is worth quoting at length:

> To transform something... is to cause a metamorphosis in form or structure, a change in the very condition or nature of a thing, a change into another substance, a radical change in outward form or inner character, as when a frog is transformed into a prince... It is change of this breadth and depth that is fostered by transforming leadership... Quantitative changes are not enough; they must be qualitative too. All this does not mean total change, which is impossible in human life. It does mean alterations so comprehensive and pervasive, and perhaps accelerated, that new cultures and value systems take the places of the old.[29]

"Mr. Gorbachev, Tear Down This Wall!" In his efforts to bring the world closer to peace, Ronald Reagan spoke his mind about the Soviet Union. Calling it an "evil empire," and demanding that its leader dismantle the Berlin Wall, Reagan displayed transformational leadership, that is leadership that motivates based on shared values and inspirational motivation.

The Prince of Transformational Leadership?

Transactional Leadership. ***In transactional leadership, on the other hand, an exchange takes place between leader and follower.*** A leader might, for example, provide rewards to an employee who meets certain deadlines or incentives. Again, one expert's definition is worth quoting at length:

> Transactional leadership [is]... the basic, daily stuff of politics, the pursuit of change in measured and often reluctant doses. The transactional leader functioned as a broker and, especially when the stakes were low, his role could be relatively minor, even automatic... [transactional leadership is] "give-and-take" leadership.[30]

FACTORS OF TRANSFORMATIONAL LEADERSHIP

Seven factors help distinguish the difference between transformational, transactional, and laissez-faire or non-leadership. Let's consider each, beginning with the four factors of transformational leadership.

IDEALIZED INFLUENCE

OBJECTIVE:
18. Explain the meaning of "idealized influence."

Leaders who succeed as role models inspire their followers to be like them. Followers choose to trust and follow these leaders largely because of the leader's ideals. In other words, ***idealized influence refers to the leader's principles and standards having the power to attract followers.***[31] Not surprisingly then, idealized influence is a key component of transformational leadership.

Leaders who practice idealized influence often display high levels of moral and ethical conduct and can be trusted to do "the right thing." In history, leaders like Gandhi exemplified this leadership factor. Gandhi led by example. He endured hunger strikes and served the lowest caste of people in India, the "untouchables." He practiced non-violent methods of protest as well and encouraged others. He wasn't just delivering his message from behind a podium, but was out in the streets, living his message.

You can be this type of leader with a flight of cadets you lead. If you are teaching flight-line management, you can order your cadets to go to the flight line and do A, B and C. Or you can go to the flight line and show the cadets how to work hard and effectively. You will likely win the loyalty of your followers by working alongside them.

A Hero to Heroes

Mohandas K. Gandhi was a legendary leader who won idependence for India by using non-violent means to expel the British.

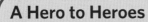
Gandhi

In our own nation, Martin Luther King Jr. admired Gandhi's philosophy and fashioned the U.S. civil rights movement along the same principles of non-violence that were espoused by Gandhi.

King

Therefore, Gandhi can be said to have had an idealized influence upon King and countless other reformers in the non-violent tradition.

INSPIRATIONAL MOTIVATION

OBJECTIVE:
19. Explain the meaning of "inspirational motivation."

In addition to idealized influence, another aspect of transforming leadership is *inspirational motivation, which describes leaders who "communicate high expectations to followers, inspiring them to become committed to and a part of the shared vision in an organization."*[31] Inspirational motivation promotes team spirit. It is the special quality of leadership that helps the team overcome impossible odds.

Consider England's King Henry V and his battle at Agincourt in France, as recounted by Shakespeare (below). The English were outnumbered nearly 5-1, though accounts of actual battlefield numbers vary.

The English used great skill with the long bow in overcoming their underdog status and winning victory, but Shakespeare also suggests that the king used inspirational motivation to inspire his troops. Yes, England's army is small and outnumbered, but that means the troops gain that much more honor. And though Henry is their king, he raises his troops to his level by calling the peasant soldiers "brothers." Inspirational leadership is what brings the English to victory.

INSPIRE THOSE YOU LEAD

Inspiring your team can be a challenge. Try using some of these ideas to become an inspirational leader.

Motivate. During drill, use cadence calls, yells, etc. Encourage your teamates to beat their personal best on tests or PT events. Try cheering on your cadets during PT.

Challenge. Organize flight challenges and create goals that require the team to succeed.

Team Mentality. Value your team's input by including them in the goal making process. Create goals that include the whole team.

Educate & Inform. Encourage your teamates to learn. Proudly teach them about our history.

Be Involved. Take a personal interest in your cadets. An active interest can help you understand what motivates members of your team and make them see that you care about their success.

Set the Example. Act as a role model of the behavior you want to see.

Celebrate Success. Reward your team with words of praise and appropriate tangible rewards. Pass on praise that you recieve about your team's work.

THE ORIGINAL BAND OF BROTHERS

King Henry V's speech to his outnumbered English Army, as told by Shakespeare, is a classic example of inspirational leadership:

We few, we happy few, we band of brothers;
For he today who sheds his blood with me
Shall be my brother; be he ne'er so vile,
This day shall gentle his condition.
And gentlemen in England now a-bed
Shall think themselves accursed they were not here,
And hold their manhoods cheap, whiles any speaks,
Who faught with us upon Saint Crispin's Day!

Henry V, Act IV, Scene 3

Photo: Phil Scarsbrook, Alabama Shakespeare Festival's 1994 *Henry V*. Foreground: Ray Chambers (Henry V).
Figures left rear include: John McWilliams, Chris Mixon, Roger Forbes, and Richard Elmore; figures right : Richard Farrell and Kurt Kingsley.

INTELLECTUAL STIMULATION

OBJECTIVE:
20. Explain the meaning of "intellectual stimulation."

Transformational leaders want to make their people smart. They value intellectual stimulation and want the team environment to be mentally and academically challenging. Instead of wanting to lead a group of robots or "yes" men, transformational leaders will encourage the team members to challenge their own beliefs as well as the leader's and the organization's as a whole. **On a team run by a transformational leader, everyone should always be learning.**[32]

In an intellectually stimulating environment, followers might discover a problem that is slowing the team down and propose a solution that rockets the organization to a new level.

To better understand how leaders can value or not value intellectual stimulation, consider two approaches to being a math tutor. On one hand, the tutor might work out the problems in front of you – basically solving them for you. The advantage here is that you finish your homework quickly, but do you really learn anything? Probably not. A better and more intellectually stimulating approach would be one where the tutors shows you how to solve a few problems, watches you try to figure some out on your own, and guides you through the tough spots. As the old saying goes, "feed a man a fish and he eats for a day, but teach a man how to fish and he'll eat for a lifetime."

TODAY'S ORGANIZATIONS VALUE INTELLECTUAL STIMULATION

Most organizations today try to recruit new employees or members by touting their culture as one that values intellectual stimulation. Here's what a few organizations have to say on the subject:

Google "When we encourage Googlers to express themselves, we really mean it... Intellectual curiosity and passionate perspectives drive our policies, our work environment, our perks and our profits."

"[At Apple,] there's plenty of open space — and open minds. Collaboration. And of course, innovation. We also have a shared obsession with getting every last detail right. Leave your neckties, bring your ideas."

LOCKHEED MARTIN "Lockheed Martin is at heart a company of inventors. We are not satisfied unless we devise a new solution – something smarter, something better. That way of thinking is also core to [our] men and women."

"The Learning Never Stops. Education is more than career training — it's the essential framework of your success in the Air Force. You'll be able to enhance your education throughout your career."

 "...and advance my education and training rapidly to be of service to my community, state, and nation..."

INDIVIDUALIZED CONSIDERATION

OBJECTIVE:
21. Explain the meaning of "individualized consideration."

As we discussed earlier (page 75), everyone is the same, and yet everyone is unique. Individualized consideration is a factor in transformational leadership that highlights that fact. Specifically, ***individualized consideration describes leaders who are supportive of followers, listen closely to them, and acknowledge their unique needs.*** These leaders coach and advise each follower on an individual basis to help them reach their full potential.[33]

Going back to Maslow (page 89), we know that everyone has slightly different needs at different times and in different facets of their lives. Leaders would be foolish to ignore what is unique about each individual and each situation.[34] Again, Coach Bear Bryant's remark that treating everyone the same is actually unfair (page 19) speaks to this point.

Anyone who has an older brother or sister understands the basics of individualized consideration. Suppose your older brother is smart or athletic or whatever the case may be. It'll be tempting for those who know him to treat you in a way that assumes you're just like him, even though you're two entirely different people. ***Transformational leadership recognizes this fact by insisting that leaders respect the individuality of each follower in helping them reach what Maslow called "self-actualization."***

Walking Blind from Georgia to Maine

Bill Irwin is blind. What can he do? Some people might assume that due to his blindness he can't even fulfill his own basic needs without help.

But Bill Irwin is anything but helpless. He is the only blind person to have hiked the 2,175 mile Appalachian Trail, a trek that countless sighted people attempt and fail.

Individualized consideration means leaders aren't supposed to make assumptions about people. In Bill Irwin's case, you'd be overlooking immense courage and strength of will. How many leaders miss opportunities by not respecting the unique talents and needs of the individual?

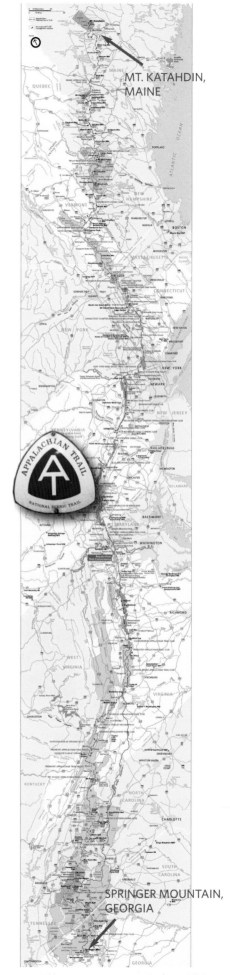

MT. KATAHDIN, MAINE

APPALACHIAN TRAIL 2,175 MILES

SPRINGER MOUNTAIN, GEORGIA

FACTORS OF TRANSACTIONAL LEADERSHIP

You see how powerful transformational leadership can be. You can work to change your followers for the better, improve their confidence and skills, and make a huge difference in their lives.

While transactional leadership has its merits, experts tend to speak less favorably of its impact than transformational leadership. *Transactional leadership, rather than focusing on lifting followers to a new level, is an exchange or transfer between leaders and followers.* Let's consider two factors associated with transactional leadership.

CONTINGENT REWARD

OBJECTIVE:
22. Explain the meaning of "contingent reward."

Contingent.
occurring only when certain
conditions are met

Contingent Reward. As a cadet, you're told that if you pass your tests and work hard, you'll receive a promotion. In the adult workforce, employees are told to work hard and they'll receive a paycheck. Contingent rewards work to some extent – employees come to work every day, after all. But relying too heavily on contingent rewards can backfire. People end up chasing the paycheck or the promotion or the carrot, versus focusing on doing the best job they can or becoming the best person they can be.

The fifth leadership factor is contingent reward, the first of two transactional leadership factors. *A contingent reward describes an interaction between leader and follower in which the follower's effort is exchanged for rewards.*[35]

In this type of leadership, the leader and follower agree from the beginning what is required of the follower and how his efforts will be recognized.[36] For instance, as a CAP cadet, you are promoted to a higher rank (your reward) after completing a number of agreed-upon requirements. (See also the section on classical conditioning on pages 91-92.)

Of course, transactional leadership takes place in much larger and often more complex settings as well, such as in American politics, a setting that one expert, James MacGregor Burns, focused on in his book *Transforming Leadership.*

Transactional leadership often takes the form of an ignoble type of deal-making that can occur between politicians who are trying to win votes or favors from opponents. "The transactional leader function[s] as a broker and, especially when the stakes [are] low, his role [can] be relatively minor."[37]

On the other hand, in transformational leadership, leaders change followers into new creations. "By pursuing transformational change, people can transform themselves."[38]

MANAGEMENT-BY-EXCEPTION (MBE)

OBJECTIVE:
23. Explain the concept of management-by-exception.

The sixth leadership factor is also more of an exchange between leader and follower than an uplifting of the follower by the leader. It takes two forms: active management-by-exception and passive management-by-exception.

Active MBE. *In active MBE, a leader watches followers closely to observe mistakes and violations of rules, then corrects the wrongs.*[39] Like a superhero flying overhead, the leader spots something the follower does wrong, swoops down to the rescue, and then returns to his perch high above the fray.

Let's go back to the flight line, where cadets are learning to direct incoming and outgoing CAP Cessnas and to communicate with pilots. A cadet observing the trainees would watch the training closely and correct the learners when they make a mistake.

In this leadership style, learning is fostered when the experienced cadet corrects errors and steers younger cadets in the right direction.

Management by Exception
What's a good example of MBE? The uniform inspection. The inspecting officer checks cadets' uniforms for mistakes and (in one way or another) ensures the problems get fixed.

Passive MBE. In passive MBE, a leader intervenes well after mistakes or problems have arisen, but never provides feedback.[40]

Returning to the flight line example above, imagine that a month after directing aircraft on the tarmac your commander surprises you with a list of mistakes you had made. It would be too late for you to fix those problems, and without meaningful feedback, you're unlikely to learn how to avoid the mistakes next time. Passive MBE fails to develop your potential.

Final Word on MBE. Whether in active or passive form, MBE is an example of transactional leadership because the leader reacts to the follower's performance, as if trading action for action. What the follower does or doesn't do triggers the leader's response.

> "Management by Exception is an example of transactional leadership."

FACTORS OF NON-LEADERSHIP

OBJECTIVE:

24. Explain the concept of laissez-faire leadership.

A final leadership factor, laissez-faire, completes the continuum that began with transformational leadership on one end and concludes with situations in which absolutely no leadership takes place at all.[41]

The word "laissez-faire" is a French word that refers to a "hands-off" or "let things ride" approach in which the leader puts off making decisions, provides no feedback and goes to little trouble to meet the needs of followers.[42]

However, laissez-faire does have its proponents. They argue that a leader is wise to allow highly-capable and dedicated followers to do their work without interference. Suppose you were given responsibility for a dining facility on an Air Force base only one hour before 1,000 airmen arrive for lunch. You see the kitchen staff working hard and efficiently. Perhaps the best approach is to stay out of their way. Laissez-faire may not count as genuine leadership, but proponents argue that it can be a responsible course for a leader.

In the end, a leader's approach to leadership is likely to have a direct impact on his or her power. In other words, the leader's employment of transformational and/or transactional leadership – particularly the leader's effectiveness with these two types of leadership – will dictate how much ability the leader enjoys to influence change.

POWER: ONLY NIXON COULD GO TO CHINA

For centuries, China kept to itself. But because China is one of the world's largest nations, doesn't it make sense for the U.S. to maintain diplomatic relations with the Chinese?

That's easier said than done, for China is communist. Therefore, most politicians were unwilling to approach the Chinese for fear they'd appear to be bowing to a communist adversary.

Enter President Richard Nixon, well known for his strongly

anti-communist views. Everyone knew Nixon would never allow a communist nation to push the U.S. around. But Nixon also believed in diplomacy and wanted to develop good relations with China.

So in 1972 Nixon visited Beijing and met with the Chinese government. Political leaders from around the globe hailed the event as a huge step forward for peace. The saying, "Only Nixon could go to China" was born.

Why did Nixon succeed? He had power. Nixon used the power of his reputation as a strong anti-communist to engage the Chinese. And none of his political opponents back home could dare question Nixon's motives.

Nixon knew how to use his power to achieve his foreign policy goals.

POWER

OBJECTIVES:

25. Define "power."
26. Describe five types of power.
27. Discuss the six stages of leadership and personal power.

A discussion of leadership is nearly purposeless without also considering the subject of power. ***Power is the ability of one person to influence another.***[43] Since a leader is someone who influences others, the leader must have power to succeed.

Legitimate Power

However, the term can be easily misunderstood. When we think of power, we sometimes might picture a leader like Hitler who was ruthless in pursuing his goals. But not only is this an over-simplification, but rather there are many types of power.

TYPES OF POWER

For our purposes, we'll focus on five types of power that are divided into two main categories: position powers and personal powers.[44] Position powers are those that people have based on their position, while personal powers are specific to the individual's knowledge and personality, not their title.

Reward Power

Position Power. Among position powers are ***legitimate power,*** in which others obey leaders because of the legitimacy of the position of the leader; ***reward power***, in which followers comply because they desire rewards that their leader can confer; and ***coercive power,*** in which followers obey because they fear punishment.[45]

Coercive Power

Personal Power. Personal power includes ***expert power***, which comes from an individual's technical knowledge, and ***referent power,*** which is conferred upon leaders when followers like and respect them.[46]

IMPLICATIONS FOR LEADERS

The most influential leaders have both position *and* personal power. They can draw upon the power of their job title, ability to reward people, potential to fire or intimidate, their immense knowledge and expertise, and the deep respect their followers have for them. The more sources of power available to a leader, the stronger and more effective he or she can be.

Expert Power

Referent Power

LEADERSHIP & POWER

According to author Janet Hagberg, the way in which leaders guide others depends on their level of personal power. In her book *Real Power*, she describes six stages of leadership and personal power.

Stage One: Domination and Force

Characteristics: In this stage, leaders depend on force and/or domination to get followers to obey. They induce fear in others because they are fearful themselves. Followers can question nothing the leader says or does.

Problem: Followers will likely rebel or mutiny after excessive Stage One leadership.

Stage Two: Seduction and Deal-Making

Characteristics: Leaders who employ the second level of personal power don't use force; instead, they are crafty. They give some reward to followers in exchange for behavior that pleases the leader. A leader might promise an off day to an employee who spends extra time on a special project.

Problem: Leaders who rely heavily on seduction and deal-making lose the trust of followers when they don't follow through on promises they make.

Stage Three: Personal Persuasion and Charisma

Characteristics: Leaders at Stage Three inspire others through their personality and charisma. Winning is the bottom line for these leaders, and they will flatter followers but also embarrass them, essentially doing anything possible to get others to discuss and agree to follow the leader's goals.

Problem: All followers will not measure success like the leader and their motivation to follow him or her will decrease.

Stage Four: Integrity and Trust

Characteristics: Leaders who employ the fourth stage of personal power try to the "right" thing; they would rather ensure that work is of high quality than to rush to complete it hastily. Followers trust them because of their honesty. These leaders aren't concerned with their own success but the entire team's performance.

Problem: Stage Four leaders can actually be hampered by their organization, which may prefer leadership that more closely resembles Stage Two and Three.

Stage Five: Empowerment

Characteristics: These leaders are called servant leaders because they focus on supporting, encouraging and loving others to bring out the best in their followers. They give to others and do not expect to be rewarded. Their goal is serve others, the direct opposite of leaders at Stage Three, who are self-serving.

Problem: Many people who have a Stage Three mentality can't comprehend the attitude or vision of these leaders.

Stage Six: Wisdom

Characteristics: People at Stage Six leaders are selfless and do not desire to lead. They still might be ideal leaders because they have achieved an inner peace that is evident in their stability and their ability to face challenges head-on. They do not fear losing anything or everything.

Problem: Some people may be uncomfortable with those at Stage Six because someone at this level of maturity often expects others to share their no-fear-of-losing value system, and others may be very uncomfortable with this philosophy.

1. Leads by force. Inspires fear.

2. Leads by seduction & deals. Inspires dependency.

3. Leads by personal persuasion. Inspires a winning attitude.

4. Leads by modeling integrity. Inspires hope.

5. Leads by empowering others. Inspires love and service.

6. Leads by being wise. Inspires inner peace.

Leadership & Personal Power

BUILDING A LEARNING ORGANIZATION

OBJECTIVES:

28. Define "learning organization."

29. Identify five disciplines one must understand to build a learning organization.

Can an organization learn? In chapter 5 we considered how individuals learn. Now, our question turns to learning at the organizational level. ***If an individual progresses and develops through learning, can an organization grow and mature through learning?*** The answer, according to experts on learning organizations, especially Peter M. Senge (pronounced Seng-eé), is a resounding "yes."

Learning organizations are places where people are continually learning together. In a learning organization, "people continually expand their capacity to create the results they truly desire, new and expansive patterns of thinking are nurtured, [and] collective aspiration is set free."[47] But how does the organization learn?

> **"In a learning organization, people are continually learning together."**

Senge contends there are "three core learning capabilities of teams" that are vital to creating a successful learning organization. To illustrate them, Senge uses a three-legged stool, for without any leg, the learning organization cannot stand.[48]

Core Learning Capabilities. The first leg of the stool is Aspiration, which includes personal mastery and shared vision. The second leg is Reflective Conversation, which involves mental models and dialogue, and the third leg is Understanding Complexity, the essence of systems thinking.

Core Learning Capabilities for the Team

Aspiration — Personal Mastery, Shared Vision

Reflective Conversation — Mental Models, Dialogue

Understanding Complexity — Systems Thinking

These three core learning capabilities (the three legs of the stool) illustrate these five learning disciplines, the fifth of which is systems thinking. An understanding of systems thinking, Senge argues, underlies all five learning disciplines.

SYSTEMS THINKING

OBJECTIVE:

30. Define "system."

31. Explain how an organization is a system.

The first step to creating a learning organization is to recognize that the organization is a large system that encompasses many smaller systems. A system, according to

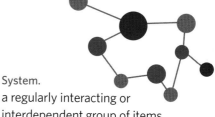

System.
a regularly interacting or interdependent group of items forming a unified whole.

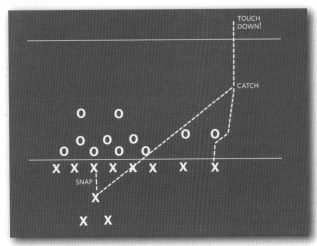

System in Play
A typical football plays shows the interaction between parts of a whole that work together.

Webster, is a "regularly interacting or interdependent group of items forming a unified whole," and, put another way, "a group of interacting bodies under the influence of related forces."[49]

Take a football team, for instance. The players regularly interact on a daily, weekly and monthly basis to perfect their strategies and to play in regularly held games. Each player depends on the other; the quarterback needs a lineman to snap the ball, and speedy running backs need huge linemen to block for them. Each player has a specific role, or mission – whether to block, or to throw the ball, or to catch the ball – but all for a unified whole. The team then is a system.

In business as well, an organization succeeds because of the interaction and interplay of many different groups of people. There's yourself. There's your immediate group; for instance, the flight of cadets that you mentor and teach. And there are more parts to your organization: the senior members of your squadron, your wing, and even National Headquarters. Your organization is a system, a large and extensive group of smaller, interacting systems.

There's a saying about disorganization: that the right hand doesn't know what the left is doing. Often organizations exist in such a way, with each department doing its own thing, not really aware of the effects of their actions on another part of the organization.

UNDERSTANDING SYSTEMS IS VITAL

If a system is dysfunctional, it can implode and even die. If it's part of a larger system, the bigger system, namely your entire organization, also could fail. Let's look at some examples of how not understanding a system can be detrimental, such as with an ecological system that contains numerous and various wildlife.

Case Study: Africa's Leopards & Baboons. Years ago, a group of farmers in Africa complained that leopards were killing their cattle and dogs. So the immediate solution was to have the government exterminate all the leopards.[50] Years later, baboons became the problem animal because leopards, the main predator of the baboon, were now gone. Leopards had been helping limit the baboon population while occasionally, of course, irritating farmers. The baboon population now skyrocketed, and since there was not enough natural food for the excess number of baboons, the baboons preyed on crops. So the baboon problem might never have occurred if farmers had left the leopards alone and used other means to ensure protection of their farms. *The farmers didn't understand the ecological system in which their families had likely lived for centuries; instead of seeing the full system, they saw but one part.*

The system you volunteer or work in is similar, believe it or not. Tweaking or adjusting one part of your organization may affect another part of your system.

Case Study: CAP's Search for Steve Fossett. One could see systems at work in Civil Air Patrol's search for legendary aviator and adventurer Steve Fossett. The Steve Fossett search showcased CAP's emergency services volunteers, who flew more than 600 flights and 1,700 hours after Fossett went missing in the Nevada desert. Meanwhile, on the ground, CAP public affairs officers in Nevada and at National Headquarters continuously sent news releases on CAP efforts to newspapers and magazines, not just in the United States but around the world. As a result of the media blitz, people worldwide learned about Civil Air Patrol, its members and its capabilities. Especially in the United States, it is likely that some private pilots may have joined CAP after learning that they, too, could help their community by flying voluntarily with CAP.

But imagine if you took away CAP's emergency services mission. Public affairs could still write about other CAP missions, but the drama of emergency services, an attraction or many prospective CAP senior members especially, would not exist.

What if you didn't have a public affairs team? Untold miracles could be performed by CAP aircrews and ground teams, and perhaps no newspapers and magazines would ever be informed of the amazing efforts.

Understanding the whole of a system, one can then focus on four other crucial steps to building a learning organization: *personal mastery, shared vision, team learning and mental models.*

SYSTEMS IGNORANCE

African farmers grow frustrated by the leopards that eat their livestock. In response, the farmers kill the leopards.

But then the baboon population, which had never been a problem, grows out of control. The leopards had been hunting the baboons and keeping their numbers down. But now that the farmers "solved" their leopard problem, the new baboon problem emerges.

Systems thinking teaches leaders to see all the different factors bearing on a problem. In a system, you can't "solve" one problem by looking at it in isolation.

PERSONAL MASTERY

OBJECTIVES:

32. Explain the concept of personal mastery.
33. Define "personal vision."

If an organization is built of parts, and a critical number of parts are actually the people who perform the organization's work, then developing each individual to their fullest capacity should improve the company or group for which each works. ***Helping each person in an organization fully realize his potential is the thrust of personal mastery.***

Personal mastery means "approaching one's life as a creative work, living life from a creative as opposed to a reactive viewpoint."[51]

A hallmark of creative work is excellence, like the brilliant artistry in a Renoir painting, the music of Mozart, or the sculpture of Michaelangelo. The work of art created by a poet or novelist is refined and refined and refined until the smallest detail of the finished product is as brilliant as the whole.

Imagine, then, if we focus such attention on all aspects of our lives, from our families and education to our work and spirit. To clarify, ***personal mastery does not refer to "acquiring more information" just to be smarter than everyone else, but rather to increase our capacity to reach our deepest personal and professional goals.***[52]

Self Mastery & the Would-Be Astronaut

You have a dream. But you struggle to get there, so you decide to lower the expectations you set for yourself. What's even worse is that your struggles don't end there and you continue to run into resistance as you pursue those lowered standards. In turn, you lower your standards again and again.

Personal mastery is an attempt to keep focused on reaching your absolute fullest potential. It is a mastery of self, like Maslow's self-actualization (p. 89).

"I want to be an astronaut."

"... but that's too big a goal, so now I want to fly the F-22 *Raptor*..."

"... but I don't really have the grades to get into college and earn a commission, so maybe I'll just enlist in the Air Force instead..."

"... but I'm not sure I'd survive Basic Training, so maybe I'll just get a job as a baggage handler at the airport."

There is dignity in each of these careers. Even a baggage handler can take pride in his job. But it's a long way from the original dream of being an astronaut.

People whose personal mastery level is high share several common characteristics:

- A unique "sense of purpose" underscores their "visions and goals."
- They view "reality" as a friend, not a foe.
- They have learned to be flexible in coping with change rather than seeing change as a problem.
- They are extremely "inquisitive," always trying to view reality more precisely.
- They feel a connection with others and life itself.

People with high levels of personal mastery continually learn; they never "arrive" but rather they practice personal mastery their entire lives. In addition, people with high personal mastery are aware of their strengths as well as their weaknesses, and their life is a constant effort to learn and grow more.

> "People with high levels of personal mastery continually learn."

Many organizations encourage employees to seek personal mastery because if each person is committed to bettering himself, ultimately the organization will also reap rewards.[53] Your squadron or wing would be no different. *If each person you know in CAP is not committed to constantly improving himself, how can your organization grow?*

PURPOSE & VISION

Two key parts of your personal mastery are your purpose and personal vision.

Purpose. *Purpose is a person's "sense of why he or she is alive."*[54] When someone knows his purpose – when he realizes what he has been placed on the earth to accomplish or do – then real and meaningful goal-setting can start.

Born & Bred to be a Police Officer
When a gunman opened fire without warning, Sgt Kimberly Munley of the Ft. Hood police department (with President Obama, above), sprang into action. Although shot three times and small in stature - friends call her "mighty mouse" - she heroically returned fire, putting down the shooter. A fellow officer explains, "[Sgt Munley] was born and bred to be a police officer." Believing you were "born and bred" for a certain job is an example of *purpose*, knowing the reason why you are alive.

Having a purpose directly affects one's attitude and outlook on life and is synonymous with genuine caring. One expert's view is worth quoting at length:

> When people genuinely care, they are naturally committed. They are doing what they truly want to do. They are full of energy and enthusiasm. They persevere, even in the face of frustration and setbacks, because what they are doing is what they must do. It is their work.[55]

Supportiveness
Caring goes hand-in-hand with having a purpose.

Personal Vision. Going hand-in-hand with purpose is the importance of having a *personal vision, a "specific destination, a picture of a desired future."*[56]

For a CAP cadet, his or her vision might be to one day earn the Spaatz Award or to solo in a Cessna aircraft. These are specific, concrete goals, but they may not be accomplished without a sense of purpose.

"Purpose is 'being the best I can be, 'excellence.' Vision is breaking the four-minute mile."[63] In other words, a sense of purpose encourages you to accomplish specific goals.[57]

In short, *having purpose and vision increases our enthusiasm and productivity because we have discovered our reason for being,* we have specific goals we would like to accomplish, and we are motivated to take the necessary steps to reach these goals.

If each member of an organization is enthusiastic and productive, ultimately, the entire organization's productivity and its success will increase.

SHARED VISION

OBJECTIVE:
34. Explain the concept of shared vision.

When each person within in organization establishes a personal vision, suddenly the organization nears a shared vision. According to one expert, "Shared visions emerge from personal visions ... In this sense, personal mastery is the bedrock for developing shared visions."[58]

Just as people are transformed when they realize their personal visions, so organizations suddenly can become places of inspiration and energy when personal visions come together to create shared visions. Even if the visions vary a good bit at the start, the members of the organization become bound by goals that everyone shares.

"What do we want to create?" those with shared vision ask.[59] You may have witnessed shared visions, or it is possible you are currently part of an organization whose members have shared vision.

Imagine a group of people realizing it wants to form a CAP squadron. They share the vision or picture of what their squadron will look like. This vision helps them set their goals and plans in motion.

But even in an already-existing organization, a shared vision can help accomplish the group's goals. The presence of a unified vision benefits the entire team. Again, the words of one expert deserve special attention:

The Final Frontier
Shared visions paved the way for men to walk on the moon.

> A shared vision ... uplifts people's aspirations. Work becomes part of pursuing a larger purpose embodied in the organization's products or services – accelerating learning through personal computers, bringing the world into communication through universal telephone service, or promoting freedom of movement through the personal automobile. The larger purpose can also be embodied in the style, climate, and spirit of the organization.[60]

Courage. ***Shared visions exhilarate people and give them courage.*** In 1961, for instance, President John F. Kennedy expressed a vision that had been growing among many in America's space program: to land a man on the moon by the end of the decade (see volume 1, pages 70-71).[61] The vision bound NASA members until in 1969, Neil Armstrong and Buzz Aldrin walked upon the moon.[62]

Risk-Taking. Putting people in space entails a degree of danger, but ***risk is a hallmark of shared visions.*** "Shared visions foster risk taking and experimentation. ... Everything is an experiment, but there is no ambiguity. ... People aren't saying 'Give me a guarantee that it will work.' Everybody knows that there is no guarantee. But the people are committed nonetheless."[63]

> **"Risk is a hallmark of shared visions."**

It's inspiring to witness and be a part of your team developing into a learning organization. To reach that goal, first, individuals focus on personal mastery and develop personal visions; then, each person begins to work together on shared visions; and then the organization becomes ready for our next topic, team learning.

TEAM LEARNING

Many of us have heard the sayings, "Great minds think alike," and, "Two heads are better than one." They are such over-used phrases, but the ideas they express are at the heart of team learning. And what do the words "minds" and "heads" have in common? It's the plural "s," indicating more than one person is involved in the learning process.

Yikes!
Your team wants to go in one direction, but your people are scattered all over. Your team needs alignment. Only then will everyone be moving toward the same goals.

Alignment. When we first transitioned from personal mastery to shared vision, there was great potential for a the team members' differing views to collide. (Have you ever tried to get ten friends to agree on what kind of pizza to order?) Therefore, ***the key to successful team learning begins with what the experts call "alignment."***[64] If a team is not aligned, or moving in the same direction, then it will look like the mess of arrows shown at left.

Fragmented teams waste energy because their members are misaligned. But when a team becomes aligned, "a resonance or synergy develops. ... There is commonality of purpose, a shared vision, and [an] understanding of how to complement one another's efforts."[65]

Synergy.
the idea that teams working together can achieve more than each individual could on his own (see volume 1, page 48)

For example, imagine that you've set out to form a model rocketry program at your squadron, but you and fellow cadet leaders have three different plans about how to proceed. Will you ever start a worthwhile, long-lasting program out of three unaligned visions? Probably not. The team is misaligned.

Team Learning.
When trying to solve a team leadership problem, the principle of insight counts. Having multiple minds at work is better than just one.

Three Dimensions of Team Learning. In team learning, in addition to the importance of alignment, there are three critical dimensions:

- insightful thinking about complex issues,
- innovative, coordinated action, and
- the role of team members on other teams.[66]

Insight. ***To think insightfully means to take advantage of the power of many minds to be more intelligent than one mind.*** While this may seem to be a given, in some organizations, team members do not acknowledge this reality.

Innovation. In addition to insightful thought, team learning also requires innovative, coordinated action. In other words, **the team needs to welcome new ideas and to work together so those ideas contribute to the team's success.** In most team sports, for example, the best teams show spontaneous yet coordinated play.[67] That is, they are innovative and adjust to what's happening on the field – "oh no, the other team is blitzing, says the quarterback, I better run with the ball!" – while continuing to work together to reach a common goal.

Roles of Team Members. Team learning require insight and synergy and highlights the important roles of the other members. That is, **learning teams that work in separate and even the same departments within an organization should help encourage each other to work in cooperation.**[68]

With these foundations in place, two of the most important ways that learning teams can reach their full potential is through dialogue and discussion.

DIALOGUE AND DISCUSSION

OBJECTIVE:
37. Explain the difference between dialogue and discussion.

Dialogue & Discussion
As a leader, it's not enough to say you value dialogue and discussion. You have to be a great listener and really push the team to sound off with their ideas.

To better understand team learning and how leaders use it to build learning organizations, it's worth taking a moment to distinguish between dialogues and discussions. Teams use dialogue and discussion as a way to learn and grow.

Dialogue. **Much like brainstorming, dialogue presents an opportunity for team members to freely and creatively explore complex issues.**[69] During dialogue, team members listen carefully as each member shares his or her points of view. The goal is to exceed any one person's understanding. That is, through dialogue the team becomes smarter than any single individual. And during the dialogue, anything can be mentioned. "The result is a free exploration that brings to the surface the full depth of people's experience and thought, and yet can move beyond their individual views."[70]

Discussion. While still involving interaction between team members, discussion is slightly different. **Through discussion, team members present differing views and defend them in a search for the best possible solution.**[71]

ESCAPE FROM OLD THINKING

Teams learn when they recognize what mental models had been control how they thought about a topic. If a team can escape from tired old mental models, it can learn and innovate.

B-2 Spirit. Who says an airplane needs a normal tail?

Hulu.com. Who says you have to be home at the right time to watch your favorite TV show?

MENTAL MODELS

OBJECTIVE:

38. Define "mental models."

Organizations whose members practice personal mastery and then turn their personal visions into shared visions can enjoy great success. The primary hindrance comes from mental models.

Mental models are "deeply held internal images of how the world works, images that limit us to familiar ways of thinking and acting."[72] They can be basic generalizations like, "people are untrustworthy," or they can be complicated theories, like the assumptions one might make about why a family's members relate with one another in a particular way.[73]

Since this topic is not very commonplace, perhaps even in discussions of leadership, let's look at an example. Chapter 6 focused partly on prejudice and hatred; prejudice is an example of a mental model. Consider a foreigner living in a brand-new country. He marries into a family that does not accept him because of his nationality. No matter what this person does, the family detests him, will not speak to him, and makes false claims about him. But there is nothing wrong with the foreigner. He is a good person doing his best to live according to the laws of his adopted country. But his new family does not accept him because it has been programmed to think, over many generations, "Foreigners from that country are bad. They are rude, dishonest and untrustworthy."

Senge points out that mental models are active. That is, *mental models cause us to act a certain way based on our assumptions, and they "shape our perceptions," meaning the way we perceive, or look at, things.*[74]

The iPad. Who says that a computer has to have a regular keyboard?

Gliders. Who says cadets can't fly?

USING MENTAL MODELS TO SUCCEED

OBJECTIVES:

39. Explain the difference between an espoused theory and a theory-in-use.
40. Explain why leaders may make leaps of abstraction.

But in the same way mental models can lead to negative actions, **teams also can use mental models to succeed.** Developing the abilities to reflect and inquire can help one work with mental models more fruitfully.[75]

Skills in reflection come from slowing down our thought processes so we can increase our awareness of how we construct mental models and how they affect our actions.[76] Inquiry skills involve how we relate with others, face-to-face, especially over complex issues.[77]

We can improve our understanding of the complexity of these processes by realizing differences between espoused theories (what we say we believe) and theories-in-use (the implied theory in what we do). We can recognize our "leaps of abstraction" (our jumps from observation to generalization), and can balance inquiry and advocacy.[78] Let's look at each.

The Espoused Theory. **An espoused theory is a line of thought that we claim to believe.** Going back to the example of a foreigner fighting prejudice from his new family, the whole family might insist it is not prejudiced against foreigners. That is their espoused theory. But, in reality, they do treat the immigrant with contempt and distrust.

The Theory-in-Use. **A theory-in-use is a line of thought representing what someone actually believes.** Again in the example of the foreigner, the family's espoused theory does not match their theory-in-use, because it judges foreigners in a different way than they claim to judge them.[79]

Recognizing that these theories don't match can help bring about positive change. Doing so requires leaders willing to work toward closing the gap between what one espouses and how one really behaves.

Espouse.
to adopt or support a cause, a belief, or a way of life

Standing in the Schoolhouse Door
In 1963, Governor George Wallace (second from left, in business suit) espoused a belief that the state of Alabama was indeed within its rights to keep blacks from attending the all-white University of Alabama. Critics charged that in reality (the theory-in-use), Alabama's segregated society was built upon whites' prejudice.

Leaps of Abstraction. *A leap of abstraction occurs when we "leap" to generalizations without testing them.*[80] Remember, "abstract" is something vague and shapeless, whereas something "concrete" is detailed and specific. A leap of abstraction can occur because our minds are not well equipped to process large amounts of detail. Hence we quickly jump to an abstract way of thinking about something. This is dangerous for leaders because you can't truly understand something if you ignore the details.

Suppose some experienced NCOs are talking with a C/SSgt who will attend her second encampment and serve as a flight sergeant. She's confused about her role and turns to higher ranking cadets for guidance. If you were to reply, "Young people today need to develop leadership skills," you'd be making a leap of abstraction. While your comment would be true, you would have unhelpfully moved the conversation away from the specific challenges of being a flight sergeant at encampment to a generic point about American youth. *Leaps of abstraction prevent teams from exploring tough problems in depth.*

GENERAL

↑

Degree of Abstraction

Young people

American teens

CAP cadets

Cadet NCOs

C/SSgt Mary Feik, who will serve as a flight sergeant at encampment

PARTICULAR

BALANCING INQUIRY AND ADVOCACY

OBJECTIVE:

41. Distinguish between advocacy and inquiry.

Inquiry occurs when you ask questions and try to gain more information so that you make the best decision possible. In contrast, *advocacy occurs when you make an argument in favor of a course of action.* When leaders over-rely on advocacy, a zero-sum game (see page 102) is apt to develop.

> "When leaders over-rely on advocacy, a zero-sum game is apt to develop."

When we meet someone to advocate for something, that is, to explain why we support something, like asking mom permission to buy a used car, your mom may have her own reasons why that's not a good idea. If both parties only bring advocacy to the table, then what may ensue is a drawn-out, back-and-forth. "I think this…" is followed by the other person saying "But I think this…" in reply. Before you know it, what started as a nice, polite conversation winds up as a heated argument.[81]

One expert calls this a "snowball effect of reinforcing advocacy" that can be eased through inquiry and actually lead to very productive and creative results on teams.[82]

Simple questions sprinkled into discussions like "What makes you think that?" "Can you illustrate your point for me?" or "Can you provide some data to back up your thoughts?" can inject some inquiry into the escalation of advocacy before it gets out of hand.

Part of the goal is to break away from a viewpoint in which there is only one winner. When inquiry and advocacy are combined, the goal is "no longer to win the argument."[83] Some practical tips:

When advocating your view:

- Make your reasoning explicit (Say how you arrived at your view and discuss the "data" upon which it is based)

- Encourage others to explore your view ("Do you see gaps in my reasoning?")

When inquiring into others' views:

- If you are making assumptions about others' views, state your assumptions clearly and acknowledge that they are assumptions.

- Don't bother asking questions if you're not genuinely interested in the others' responses.[84]

LEADERSHIP STYLES

Assuming you have a team in place, the question may arise, "How do I lead?" Well, there is no simple, ideal answer. But that's not bad news; the leader's toolkit is filled with options.

To help understand leadership better, experts offer various leadership theories, which focus on relationships between leaders and followers as well as the situations leaders find themselves in. In the following sections, we'll discuss three popular leadership theories: situational leadership, the path-goal model, and the leadership grid.

A LEADER'S VIEWPOINT

OBJECTIVES:
42. Define "task behavior."
43. Define "relationship behavior."

To help understand various leadership theories, experts have found that leaders exhibit essentially two kinds of behaviors: task and relationship behaviors.[85]

PRESS SHUFFLE *on the* **LEADERSHIP PLAYLIST**

They go about the job of leading in different ways. Maybe one is an innovator, another the take-charge sort, or a consensus-builder. Which is the right way? Most experts today say it depends on the situation.

Task behaviors involve actions that relate to how a job or project gets done directly in terms of organization of work, scheduling of work, and who will perform individual tasks.

Relationship behaviors include building morale, respect, trust, and fellowship between leaders and followers.[86]

Experts contend leaders must be able to balance these two behaviors appropriately depending on each situation. Let's look at how these behaviors play a part in the Blanchard situational theory.

SITUATIONAL LEADERSHIP

OBJECTIVES:

44. Summarize the main idea that underlies situational leadership theory.
45. Describe the authoritarian, democratic, and laissez-faire leadership styles.
46. Describe situations where each style would be appropriate.
47. Explain why some critics believe situational leadership is not an effective way to lead.

> "Every challenge is unique, so match your leadership style to the situation."

Every challenge is unique. Therefore, match your leadership style to the situation. That's the overall idea behind situational leadership. For our purposes here, we'll consider three distinct leadership styles: authoritarian, democratic, and laissez-faire.

AUTHORITARIAN

You might be a very easygoing, relaxed person, but there are times when you will need to be more assertive. *The authoritarian style is ideal when you have time limits or critical situations, or when individuals cannot respond to less direct approaches.*

For example, if you are at the scene of a car accident and a victim is about to die, then you may give specific, direct orders that in a normal situation might seem rude – "Get me a towel!" "You! Hurry, call 911!" and "Get over here now!"

Similarly, if you are training some cadets who won't listen to your polite requests to stop horsing around, then you might have to raise your tone – "Stop it! We're conducting training."

The caution with the authoritarian style is that if you overuse it, it becomes counterproductive because you are trying to motivate by instilling fear. People cannot demonstrate their full potential in an atmosphere of fear.

Note (page 145)
In the context of situational leadership, laissez-faire suggests a beneficial leadership style. It is true that in our earlier section on transactional leadership, we considered it synonymous with the nonexistence of leadership. Here, laissez-faire is seen in a more positive light.

DEMOCRATIC

The democratic leadership style (also called 'participative') involves effective listening, rational dialogue and discussion, and consideration of others. This style is ideal for situations in which you want all of your followers to take responsibility for equal parts of achieving a common goal.

In many ways, the democratic style combines the ideas we've related earlier in this chapter, such as strengthening emotional intelligence and building learning organizations. That is, *democratic leadership calls for breaking down barriers between followers and the leader.*

For example, a cadet commander may use a democratic approach by gathering all the cadets together and inviting everyone to have a say in what the squadron's goals will be for the coming year. Leaders who use a democratic style would reason, "Why shouldn't the cadets' own interests be used in deciding whether we'll focus on drill team, emergency services, model rocketry, and the like?"

The democratic style teaches that followers must learn to care for one another, to value the viewpoints of others, and to take part in fruitful, non-combative dialogue and discussion aimed at reaching positive solutions.

LAISSEZ-FAIRE

The guiding principle of the laissez-faire style is that a well-trained and capable team should be empowered to work on its own. You can still stay and watch your team work, but no longer must you provide detailed instructions. In fact, laissez-faire leaders believe that were they to intervene in the team's work, they'd mostly just get in the way.

For example, your squadron is located in a college town and an outstanding Spaatz cadet transfers in to your unit. Knowing how capable she is, the cadet commander reasons a laissez-faire style is a good match for the situation. He asks her to take charge of training a color guard for a local Veterans' Day service. The Spaatz cadet agrees and that's it – the cadet commander walks away confident the job will get done.

You can use a laissez-faire approach when your cadets are doing their jobs smoothly in a well-coordinated and productive way. A great advantage of this style is that it allows the leader to focus on the big picture, versus the workaday concerns of the team.

AUTHORITARIAN

Welcome to the Air Force Academy! The classic example of authoritarian leadership is the military drill instructor, or here a first class cadet at the Academy, barking orders to a trainee.

DEMOCRATIC

One Man (or Woman), One Vote
America is built upon democratic values. It's no surprise then that most organizations, including CAP, are governed by a board that makes decisions by voting and other democratic means.

LAISSEZ-FAIRE

Hands-Off the Economy
Milton Friedman (here with President and Nancy Reagan) applied a laissez-faire philosophy to economics. He and Reagan argued that for the economy to perform at its best, government should intervene as little as possible.

PROS & CONS OF SITUATIONAL LEADERSHIP THEORY

Leadership styles are like tools in your toolkit. You can use an authoritarian, democratic, or laissez-faire style depending on the situation, just as you might use a wrench or a hammer depending on what you're trying to build or fix. *The situational approach gives leaders options.* But what if you're just not an authoritarian type of person? Or maybe the laissez-faire approach is incredibly tough to pull-off, even if you try extra hard? Critics charge that situational leadership fails because it asks leaders to be great in every approach to leadership. *You can't just "be yourself," say the critics, you have to play the part of the authoritarian, the democrat, or the hands-off leader.*

THE PATH-GOAL MODEL

OBJECTIVES:
48. Describe the path-goal leadership model.
49. Identify four possible leadership behaviors in the path-goal model.

> "The path-goal model aims at motivating subordinates to accomplish team goals."

The path-goal leadership model focuses on the exchange between leaders and followers.[87] It is aimed at motivating subordinates to accomplish team goals.[88]

Specifically, the goal of this leadership model is to "enhance employee performance and employee satisfaction by focusing on employee motivation."[89] Hence, the name *path-goal refers to the role of the leader to clear paths subordinates have to take in order to accomplish goals.*[90]

But path-goal is also an exchange in which leaders and followers trade guidance or support (provided by the leader) for productivity (provided by the follower) and satisfaction (gained by the follower).[91]

The primary make-up of the path-goal model includes a balance of the behaviors the leader is meant to adopt — whether it be directive, supportive, participative or achievement-oriented — as well as subordinate characteristics and task characteristics. All of the above motivates subordinates toward the anticipated goals as well as productivity.

COMPONENTS OF THE PATH-GOAL MODEL

Directive Leadership. *A leader who is directive gives followers specific instructions about the task,* including the leader's expectations of the follower and guidance on how to proceed, as well as any particular methods to use and a deadline.[92] The directive leader "sets clear standards or performance and makes the rules and regulations clear to subordinates."[93]

Supportive Leadership. *Supportive leaders ensure that the well-being and human needs of followers are met.* They are often kind and easy to approach. An emphasis on making work enjoyable for subordinates typifies this type of leader.

Participative Leadership. *This leadership behavior characterizes leaders who allow subordinates to share in decision-making.*[94] Leaders who welcome participation consult with followers, gather their ideas and opinions and incorporate followers' feedback into decisions about how the group will move forward.[95]

Achievement-Oriented Leadership. While the leadership behaviors advocated by path-goal are weighted toward ensuring followers' needs are met, the theory also advocates leadership that encourages followers to excel and continuously improve.[96] *Achievement-oriented leaders show great confidence that followers are able to set and meet goals that are challenging.*[97] The particular behavior that the leader chooses depends on the both the task and the follower.

Task and Subordinate (or Follower) Characteristics. *The theory assumes that leaders can "correctly analyze the situation," decide on the appropriate behaviors and adjust their behavior to the situation.*[98] For instance, if the task is boring or stressful, the leader must show support and consideration to remove barriers to follower satisfaction.[99] Similarly, if the task is complicated or brand-new, the leader must be directive.[100] Perhaps the follower may need step-by-step instructions on how to perform a particular procedure. By assisting the follower with direction, the leader once again removes barriers that could decrease the follower's motivation and satisfaction.

PATH-GOAL IN PRACTICE

The structure of the task is a critical aspect of path-goal theory. That is, once again to use the example of asking a follower to design a public relations campaign, this task would be of low structure. There would not be a clear-cut set of guidelines to follow in initiating such a task and measurement of success would be difficult. In this scenario, the satisfaction of the follower could be challenged greatly. Hence, a supportive leadership style that encourages the follower would be vital.

ON A PATH TOWARD A GOAL

According to the path-goal model, if you show the way and help your people reach the goal, you're a leader. Outlined below are three principles of path-goal:

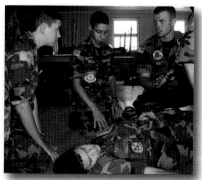

(1) Clarify the path so the team knows which way to go.
This cadet officer is explaining how to perform first aid. "Here's the job, here's the procedure, this is the standard... Any questions?"

(2) Remove roadblocks that are keeping the team from the goal.
NCOs can run a color guard, but cadets still need help. Unless they obtain flags, rifles, transportation to the event, etc., the cadets' success will be blocked. By properly equipping the team, the cadets' leaders have prepared them for success.

(3) Offer rewards along the way.
As the followers march along the path to their goals, leaders should provide encouragement and rewards. Honor Cadet awards are a good example.

TWO WAYS TO LEAD, ONE GRID

The 9,1 Leader
Leaders who have a high concern for results or task behavior focus on getting the job done. Leaders who come from this perspective would say it doesn't matter if this loadmaster is having a bad day or doesn't feel he's a real part of the team – all that matters is that he gets the job done.

The 1,9 Leader
Leaders who have a high concern for people or relationship behavior focus on taking care of their people. Leaders who come from this perspective would say that people are their biggest asset – take care of your people and they'll amaze you with their performance.

As you see, *the nature of the task affects the type of leadership required of the leader.* And yet the nature of the follower – subordinate characteristics – and not just his response to the task is important as well. For example, some employees appreciate "guidance and clear instructions," so a directive leadership approach is suitable in this instance.[101] Followers who are self-starters and prefer more independence and freedom when working a task would prefer a less directive approach and more emphasis on supportiveness.[102]

GRID THEORY OF LEADERSHIP STYLES

OBJECTIVES:
50. Discuss the goals of the leadership grid.
51. Describe five primary leadership types found on the leadership grid.

Self-deception is a major barrier to effective leadership. We are often blind to how our behavior comes across to others. When assessing our own behavior we see our *intentions* – the ideal behaviors that we espouse. Others see our *actual* behavior, and the two views are often very different! Because our behaviors are tied to core values, beliefs, and assumptions, we take them personally and often reject or dismiss challenges out of hand.

The grid theory of leadership styles was developed by Drs. Robert R. Blake and Jane S. Mouton in 1961 as an objective point of reference for exploring behaviors. *The grid theory allows people to discuss behaviors in a "disconnected" way that reduces defensiveness and judgment.* For example, you can say, "I disagree with *this style*," versus, "I disagree with *you.*"

An Easy Way to Discuss Leadership. *Grid theory also gives people a common language that can be shared.* This sharing accelerates candor development as people converge around behaviors they agree are sound and reject behaviors that are not.

Grid vs. Situational. The difference between grid and situational theories is that *Blake and Mouton believed behaviors are rooted in core beliefs, values, attitudes, and assumptions that are ingrained and remain constant in leaders across situations.* They also believed the two concerns are interdependent rather than additive; meaning *how* a person expresses a "high" concern for people, for example, depends on the interacting level of concern for results present.

INDIVIDUAL GRID STYLE SUMMARIES

By plotting coordinates on the nine-by-nine grid, 81 styles are possible. In practice, seven styles stand out. These distinct styles emerge in relationships when the two concerns interact as people work together.

The 9,1 Style: Controlling (Direct & Dominate): The 9,1-oriented person demonstrates a high concern for results with a low concern for people. He or she believes that expressing a high concern for people will diminish results. The attitude is, "People are only productive when constantly pushed."

The 1,9 Style: Accommodating (Yield & Comply): The 1,9-oriented person demonstrates a low concern for results with a high concern for people. He or she believes that expressing a high concern for results will diminish morale and therefore results. The attitude is, "Happy people are more productive."

The 5,5 Style: Status Quo (Balance & Compromise): The 5,5-oriented style is found in the middle of the grid figure with medium levels of concern for both people and results. The attitude is, "You have to balance people needs with productivity needs by using traditions, past practices, and popular trends as the basis for leadership."

The 1,1 Style: Indifferent (Evade & Elude): The 1,1 oriented style demonstrates the lowest concern for both people and results. He or she believes that expressing a high concern for either results or people will not make a difference. The attitude is, "I don't matter as an individual so why should I make any extra effort?"

The Paternalistic Style (Prescribe & Guide): Paternalism is a result of 9,1 and 1,9 joining to make a unique style. The paternalist uses him or herself as the ultimate model for everyone to follow. Loyalty is encouraged and rewarded. Disloyal followers are made miserable by isolation and increased scrutiny. The attitude is, "I know what's best and people need my guidance."

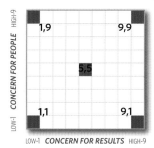

THE LEADERSHIP GRID

® reprinted courtesy of Grid International
gridinternational.com

CONCERN FOR PEOPLE HIGH-9 / LOW-1

1,9   9,9

5,5

1,1   9,1

LOW-1   *CONCERN FOR RESULTS*   HIGH-9

9,1 Controlling
Direct & Dominate
I expect results and take control by clearly stating a course of action. I enforce rules that sustain high results and do not permit deviation.

1,9 Accommodating
Yield & Comply
I support results that establish and re-inforce harmony. I generate enthusiasm by focusing on positive and pleasing aspects of work.

5,5 Status Quo
Balance & Compromise
I endorse results that are popular but caution against taking unnecessary risk. I test my opinions with others involved to assure on-going acceptability.

1,1 Indifferent
Evade & Elude
I distance myself from taking active responsibility for results to avoid getting entangled in problems. If forced, I take a passive or supportive position.

Paternalistic
Prescribe & Guide
I provide leadership by defining initiatives for myself and others. I offer praise and appreciation for support, and discourage challenges to my thinking.

Opportunistic
Exploit & Manipulate
I persuade others to support results that offer me private benefit. If they also benefit, that's even better in gaining support. I rely on whatever approach is needed to secure an advantage.

9,9 Sound
Contribute & Commit
I initiate team action in a way that invites involvement and commitment. I explore all facts and alternative views to reach a shared understanding of the best solution.

Paternalism.
Acting like a father to your followers, with the authority and wisdom that implies

The Opportunistic Style (Exploit & Manipulate): The opportunist-oriented style is the only situational style on the grid. The opportunist approaches every situation with the underlying attitude of "What's in it for me?" and then takes on whatever style is most likely to result in private advantage. The key to successful opportunism is the ability to persuade people to support selfish objectives without revealing the underlying motives. The inconsistency in the approach used makes the style difficult to identify in the short term.

The 9,9 Style: Sound (Contribute & Commit): The 9,9-oriented style is considered the ideal style and integrates a high concern for people with a high concern for results. This leader sees no contradiction between the two concerns and so builds resilient relationships that overcome challenges and strive for excellence.

IS THE GRID HELPFUL?

How do I know where my natural leadership style fits on the grid? Some leadership experts would argue that the grid's designers are biased in favor of 5,5 or 9,9 leaders, even though history shows leaders of other stripes can succeed. Situational leadership theorists would say the leadership grid is fine at describing different styles, but the key is to use the right style for the right setting. Once again, it seems leadership theory provides more questions than answers.

CONCLUSION

By now you have a better understanding of yourself, your team and leadership theories that can guide you on your leadership journey. Remember, before you can lead a team, you must lead yourself. Emotional intelligence helps you gain control of you.

As you begin to gain personal mastery through careful attention to your growth and development and to living your life as though it is "a work of art," you'll become a strong team member and leader.

By adapting transformational and transactional leadership styles that fit your individual leadership style, you can lead and even grow your team exponentially while becoming the type of leader who truly cares for and nurtures followers.

You will have a learning organization, and you will be successful as a leader.

DRILL & CEREMONIES TRAINING REQUIREMENTS

As part of your study of this chapter, you will be tested on the drill and ceremonies listed below. Ask an experienced cadet to assist you in learning about these functions. For details, see the *USAF Drill and Ceremonies Manual*, or the *Guide to Civil Air Patrol Protocol*, available at capmembers.com/drill and capmembers.com/pubs, respectively.

From the *Air Force Drill & Ceremonies Manual*, Chapter 6, Section B

Forming the Group

Dismissing the Group

From the *Air Force Drill & Ceremonies Manual*, Chapter 7, Section C

Raising & Lowering the Flag

Reveille and Retreat Ceremonies

From CAPP 3, *Guide to Civil Air Patrol Protocol*, Attachment 4

CAP Change of Command Ceremony

ENDNOTES

1. Stephen R. Covey, *Seven Habits of Highly Effective People*, (New York: Simon & Schuster, 1989), 98.

2. Hendrie Weisinger, Ph.D., *Emotional Intelligence at Work*, (San Francisco: Jossey-Bass, 1998), xvi.

3. Afsanah Nahavandi, *The Art and Science of Leadership*, (Upper Saddle River, NJ: Prentice Hall, 2003), 67.

4. Ibid, 66.

5. Daniel Goleman, *Emotional Intelligence*, (New York: Bantam Books, 1995), 47.

6. Ibid, 47.

7. Ibid, 48.

8. Ibid, 48.

9. Weisinger, 6.

10. Ibid, 6.

11. Merriam-Webster's Collegiate Dictionary, 10th ed., s.v. "self-fulfilling prophecy," 1061.

12. Weisinger, 6-7.

13. Ibid, 13.

14. Ibid, 64.

15. Ibid, 69.

16. Ibid, 69.

17. Ibid, 73.

18. Ibid, 69.

19. Goleman, 90.

20. Weisinger, 185.

21. Ibid, 185.

22. Ibid, 185.

23. Ibid, 185.

24. Goleman, 39.

25. Ibid, 118.

26. Peter G. Northouse, *Leadership Theory and Practice*, (Thousand Oaks, Calif: Sage Publications, 2001), 135.

27. Ibid, 132.

28. Ibid, 132.

29. Ibid, 176-177.

30. Ibid, 137.

31. Ibid, 138.

32. Northouse, 138.

33. Ibid, 138.

34. Ibid, 138.

35. Ibid, 139.

36. Ibid, 140.

37. Ibid, 140.

38. James MacGregor Burns, *Transforming Leadership*, (New York: Grove Press, 2003), 24.

39. Ibid, 26.

40. Northouse, 140.

41. Ibid, 140.

42. Northouse, 141.

43. Ibid, 141.

44. Ibid, 141.

45. Nahavandi, 97.

46. Nahavandi, 100-101.

47. Ibid, 100-101.

48. Ibid, 101.

49. Peter M. Senge, *The Fifth Discipline: The Art & Practice of the Learning Organization*, (New York: Doubleday, 2006), 3.

50. Ibid, xiii.

51. Merriam-Webster's Collegiate Dictionary, 10th ed., s.v. "system," 1197.

52. Michael Scott, *The Young Oxford Book of Ecology*, (New York: Oxford University Press, 1995), 9.

53. Senge, 131.

54. Ibid, 132.

55. Ibid, 133.

56. Ibid, 137.

57. Ibid, 138.

58. Ibid, 138.

59. Ibid, 139.

60. Ibid, 197.

61. Ibid, 192.

62. Ibid, 193-94.

63. Ibid, 194.

64. Ibid, 194.

65. Ibid, 195.

66. Ibid, 217.

67. Ibid, 217.

68. Ibid, 219.

69. Ibid, 219.

70. Ibid, 219.

71. Ibid, 220.

72. Ibid, 224.

73. Ibid, 163.

74. Ibid, 167.

75. Ibid, 167.

76. Ibid, 175.

77. Ibid, 167.

78. Ibid, 175.

79. Ibid, 176.

80. Ibid, 177.

81. Ibid, 178.

82. Ibid, 183.

83. Ibid, 184.

84. Ibid, 185.

85. Ibid, 186.

86. Northouse, 35.

87. Ibid, 36.

88. Nahavandi, 160.

89. Northouse, 89.

90. Ibid, 89.

91. Nahavandi, 161.

92. Ibid, 161.

93. Northouse, 92.

94. Ibid, 92.

95. Ibid, 92.

96. Ibid, 92.

97. Ibid, 92.

98. Ibid, 92.

99. Nahavandi, 161.

100. Ibid, 161.

101. Ibid, 161.

102. Ibid, 161.

103. Ibid, 161.

CHAPTER 8
EFFECTIVE COMMUNICATION

THE LEADER WHO SPEAKS AND WRITES WELL is an effective leader. When an individual shows good communication skills, others look favorably upon that person's intelligence, persuasiveness, and self-confidence. The effects of good communication skills happen to be some of the same attributes people admire in leaders. Leaders who understand that fact and continually develop an ability to speak and write effectively will make their voices heard over the din of so many others whose ideas command no attention.

CHAPTER GOALS

1. Develop an awareness of the importance of effective communication

2. Understand how to write an essay

3. Understand how to prepare and present a speech

4. Appreciate how communication skills can affect your career and life

COMMUNICATION FUNDAMENTALS

OBJECTIVE:
1. Name three purposes of communication.

You can thrive, whether writing or speaking, if you understand several foundations of successful communication. These include your purpose, knowledge of your audience, and your organization.

KNOW YOUR PURPOSE

Every speech or essay should have a specific purpose, an exact statement of what you want your audience to understand, do, or believe. In other words, why are you writing or speaking? Do you want to entertain your audience, to inform it of something you feel should be known or understood better, such as the benefits of glider flying, or to persuade audience members to change their viewpoint on how they feel about home schooling, abortion, or the success or failure of American war planning?

If you are speaking about the benefits of glider flying, for instance, your speech could start with a statement like, "Today I would like to share with you my experiences at the recent glider academy and why I feel every cadet should learn to fly gliders in addition, of course, to powered aircraft."

Your purpose could be a combination of entertaining, informing, and persuading, but most often in CAP you will be informing or teaching. Especially as a cadet NCO, you may find yourself teaching younger members about drill and ceremonies and other CAP traditions. As a flight sergeant or first sergeant, you might speak to younger cadets formally or informally on the importance of living the Core Values. If so, your purpose is to ensure that your communication ***teaches*** the Core Values to your cadets.

KNOW YOUR AUDIENCE

OBJECTIVES:

2. Define "audience."
3. Explain the importance of knowing your audience.

But equally important to your purpose is an awareness of *your audience – that is, those to whom you will speak or write*. Knowledge of your purpose and audience go hand in hand. It is your audience you will teach, or instruct, or entertain. Moreover, writers have catered to their audiences for thousands of years. Your respect for your audience carries on traditions of writers like Horace, born in 65 B.C., who believed the writer should delight his audience, teach it, or do both.[1]

Triple-threat
A speech can entertain, inform and/or persuade.

Consider this. You've got a 9 a.m. Saturday presentation to the model rocketry club. You are so excited to share your love of the hobby with your peers. To prepare, you create an excellent slideshow on basic model rocketry. It's sure to be a hit. But when you get to the meeting and begin to speak, your audience is bored stiff; all are advanced model rocketry students who already know everything you are teaching. They needed advanced concepts to stretch their minds further; instead they ended up feeling shrunk. Your slideshow was great, but it was built for a less experienced group of students. You should have analyzed your audience's needs better.

ORGANIZE YOUR IDEAS

OBJECTIVE:

4. Define "outline" and explain its importance.

First, you have to know why you are writing and speaking and to whom you are communicating. Next, whether writing or speaking, your organization is critical. *Organization refers to the way you put something together.* For instance, if you are building a model airplane, are you going to paint it, then assemble the pieces, or the reverse? The success of your finished product depends on decisions you make regarding assembly, or how you will put your communication together.

Creating an outline – a diagram that shows how your communication will be organized — is a great way to start. For instance, if you are writing an essay about CAP's Core Values, then your outline could show each value.

FOCUS PRINCIPLES *for* WRITING & SPEAKING:

Focused
Address the issue, the whole issue and nothing but the issue.

Organized
Systematically present your information and ideas.

Clear
Communicate with clarity and make each word count.

Understanding
Understanding your audience and its expectations.

Supported
Use logic and support to make your point.

For instance, your outline may look like this:

Topic: Core Values

I. Opening paragraph
   (or opening remarks for a speech)

II. Body
   A. Integrity
   B. Respect
   C. Volunteer Service
   D. Excellence

III. Conclusion
   (or closing remarks if speaking)

A misconception for those new to writing and speaking may be that the outline is difficult. Actually, it doesn't have to be. *The outline helps you to organize what you will write or say. It also helps you ensure that you don't omit vital information from your presentation and therefore helps ensure that your content leads to fulfillment of your purpose.* Once you begin writing or preparing a speech, you can literally use your outline as a checklist, crossing off items after you've included them in your essay or speech.

If you were presenting a speech about Civil Air Patrol's missions, the body could break quite easily into three paragraphs that cover Cadet Programs, Aerospace Education and Emergency Services. Other speeches or essays may be more complex – focusing on aspects of a single CAP mission, for example – but the general idea is to use the outline to organize your communication.

However you communicate – whether in speech or writing – realize that while there are similarities between the two crafts, they aren't exactly the same.

AUDIENCE ANALYSIS:
PRESIDENT FORD PARDONS THE "DRAFT DODGERS"

Hundreds, if not thousands of young men illegally refused to be drafted into the military during the Vietnam War.

It's 1974 and Gerald Ford is president. He knows that Vietnam has divided America and that the nation desperately needs to heal itself. But how? Until we can put the war behind us, Ford reasons, we won't be able to focus on any of the other tough problems facing America.

Ford's solution involves forgiveness. He decides to pardon all the so-called "draft dodgers," many of whom are living in Canada and unable to return home for fear of being arrested. Under Ford's plan, all young men who fled the draft will be forgiven if they perform two years' public service. Ford concludes that this is the best way for the country to move beyond Vietnam. He knows it will be a very controversial decision.[2]

Ford could announce his plan for amnesty at a meeting of anti-war protesters and enjoy loud and prolonged cheers. They'd love him for it.

Instead he finds the toughest possible audience: the Veterans of Foreign Wars. (A World War II veteran, Ford himself is a VFW member.) At the VFW national convention he tells those heroic, battle-tested men that he's pardoning the "draft dodgers." Of course, they are furious.[3]

Many communication experts advise leaders to avoid hostile audiences. Persuade and smooth-over your differences, they say. President Ford showed leadership by courageously telling critics of his plan, face-to-face. He didn't shirk responsibility.

FOLLOW BASIC COMMUNICATION PRINCIPLES

OBJECTIVE:

5.   Identify and describe six vital communication principles that will help your speaking and/or writing.

Whether you are writing or speaking, certain principles of composition are always relevant. To become an effective communicator, master principles such as these:

Be Clear – **Make your meaning clear by using definite, specific, concrete language.**[4]

> WEAK *CAP exposes cadets to aviation.*

> CLEAR *CAP cadets learn about aviation through orientation flights in single-engine Cessnas and gliders.*

What makes the first sentence weak and the second more clear? In the first, the verb (*exposes*) is vague. In the second, the verb (*learn*) is more concrete. Also, the increased details of the second sentence add to its clarity.

Use Familiar Words – Some speakers and writers think they will look and sound like a superstar if they use big words with more than two syllables.[5] But in speaking and writing, the goal is to communicate your message, not to impress or confuse the audience. **Use a familiar word unless a ten-dollar word is needed.**

> $10 WORDS *To effectuate change, it would behoove us to disseminate the strategic action plan.*

> FAMILIAR WORDS *Let's give everyone a copy of the plan so they know about the upcoming changes.*

But aren't ten-dollar words better than familiar words? Doesn't serious literature use ten-dollar words and avoid the familiar? Consider this famous passage that uses only simple words:

> FAMILIAR WORDS *In the beginning God created the heaven and the earth. And the earth was without form, and void; and darkness was upon the face of the deep.*

Eliminate Clutter – **Omit needless words.** A sentence should have no unnecessary words, a paragraph no unnecessary sentences.[6]

> CLUTTERED *In my opinion, the greatest moment in human history was when Neil Armstrong and Buzz Aldrin walked on the surface of the moon after landing their spacecraft in the year 1969. (31 words)*

"Does [writer William Faulkner] really think big emotions come from big words? He thinks I don't know the ten-dollar words.

I know them all right. But there are older and simpler and better words, and those are the ones I use."

ERNEST HEMINGWAY

156

EFFICIENT   *The greatest moment in human history was in 1969 when Neil Armstrong and Buzz Aldrin set foot on the moon. (20 words)*

Stay Active – Write and speak in the active voice. The active voice is usually more direct and compelling than the passive voice.[7]

PASSIVE   *My first flight in an airplane will always be remembered by me.*

ACTIVE   *I'll always remember my first flight in an airplane.*

What's the difference? **With active voice, the subject is doing the action.** Also, the main verb (*remember*) does its job without help. With passive voice, it feels like the object is doing the action, and the main verb (*remember*) needs assistance from helping verbs (*will be* remembered). Too often, passive voice is wordy and boring. Active voice has punch.

Put Statements in Positive Form.[8] Tell the reader or audience what is happening, what you believe. In other words, don't tell people only what not to do.

NEGATIVE   *Cadets are not to forget to bring their canteens and compasses to the training.*

POSITIVE   *Cadets must remember to bring their canteens and compasses to the training.*

Use Parallel Structure. Use the same grammatical form for expressions that are part of a group.

NON-PARALLEL   *The cadets, senior members, and the Air Force officers attended the wing conference.*

PARALLEL   *Cadets, senior members, and Air Force officers attended the wing conference.*

NON-PARALLEL   *I want to attend encampment and I want to attend Cadet Officer School and a flight academy.*

PARALLEL   *I want to attend encampment, Cadet Officer School, and a flight academy.*

Parallel structure comes down to consistency. Pick a style and stick with it. The reader or audience will have an easier time following your ideas.

"There is nothing to writing. All you do is sit down at a typewriter and bleed."

ERNEST HEMINGWAY

WRITING EXCELLENT ESSAYS

You can write. And you can write well. You don't have to be a world-renowned author.

If you enjoy thinking, then writing is just putting your thoughts on paper. Creativity is a gift we all can utilize and strengthen. The challenge is to have someone read our writing. It will have to be interesting and informative. The reader wants to learn something he or she didn't know. *In the modern day of hurry-up-and-go, readers will only take a few seconds to decide to continue reading before they either keep reading or quit.*

Yes, you are right. The essay you write for your teacher or the senior member in your squadron will be read and graded, regardless of its merit. But in the world of professional writing, where salaries are made and paychecks cut, your writing will only be read if a reader – the audience – finds it provocative and worth his or her time.

Ears Have it
Understanding your audience — to whom you are communicating — is vital.

THE GOAL OF WRITING

OBJECTIVE:
6. Identify the main goal of written communication.

Your main goal in writing (or speaking) is to share meaning, and, in doing so, inform, persuade or entertain. It is also important to communicate your message to your audience clearly and without distractions that can occur from wordy sentences, incorrect spelling, and grammatical errors.

To really make your writing sing, you must present a thoughtful and logical argument for the cause you advocate or the theory you propose. Essentially, you must support your claims with data and examples. That is, it's not enough to say that Civil Air Patrol benefits youths tremendously; you must communicate *what* those benefits are and *how* they help cadets.

Before you research, however, excellent writing begins by emptying your thoughts onto paper.

BRAINSTORMING

OBJECTIVE:

7. Describe "brainstorming."

It would be nice to think of writing like a Wyatt Earp scene, where he pulls his gun from his holster and shoots down the bad guy in a split second. Bang, you're done! In writing, fortunately, no one has to be shot, but the slower draw usually wins.

Before you start writing, brainstorming will be a good way for you to gather ideas. Get out some paper and write down anything that comes to mind about your subject.[9] When you're done, you'll likely have a lot of ideas on paper that you can use to support your arguments. Brainstorming helps break writer's block; in other words, it's better to write something than spend all day staring at a blank page.

Brainstorming
For more on brainstorming and other techniques of creative thinking, see chapter 5.

Writer's Block.
The feeling of frustration you get when you can't quite figure out what to write about.

MAKING AN ARGUMENT

OBJECTIVE:

8. Describe "argument," in the context of an essay's body.

There is a puzzled look on your face, perhaps. Yes, some youths might grow up thinking that arguments are only disputes with brothers and sisters over who will wash the dishes or take out the trash. But Webster defines ***arguments as "reasons given in proof or rebuttal."***[10] Additionally, Webster says a ***reason is a "statement offered in explanation or justification."***[11] In other words, when you write or speak, you must provide explanations to support your viewpoints.

Suppose you are asked to write about leadership mistakes you have made and explain what you learned from them. Since the bare minimum for an essay calls for three body paragraphs between your opening paragraph and conclusion, it will be logical to discuss three mistakes you have made or witnessed and devote a paragraph to each. In each paragraph, it will be wise for you to explain – to give reasons or arguments – why they were mistakes and, as it says above, what you learned from them.

SPELL CHEQUER

Eye halve a spelling chequer
It came with my pea sea
It plainly marques four my revue
Miss Steaks eye kin knot sea.

Eye strike a key and type a word
And weight four it two say
Weather eye am wrong oar write
It shows me strait a weigh.
As soon as a mist ache is maid
It nose bee fore two long
And eye can put the error rite
Its rarely ever wrong.

Eye have run this poem threw it
I am shore your pleased two no.
Its letter perfect in it's weigh
My chequer tolled me sew.

—Sauce Unknown

TOPIC SENTENCES

OBJECTIVE:
9. Explain the purpose of a topic sentence.

Your arguments will be the backbone of your essay and become the topic sentences of your three body paragraphs. *A topic sentence introduces the main idea of a paragraph.*

Consider an essay explaining the need for aerospace education. Let's say you brainstormed on this topic and developed three very strong ideas: (1) that aerospace education helps protect our country; (2) that aerospace education continues to nurture space exploration; and (3) that aerospace education inspires students to consider employment in a field that is vital to America. Voila! The ideas you developed can now become great topic sentences.

For instance, your first body paragraph will begin with a sentence on defense of our country. It might say:

> *Aerospace education is vital to the defense of our country.*

Then you will follow that topic sentence with a handful of sentences that support that main idea. Your paragraph may look like this:

> *Aerospace education is vital to the defense of our country. First, it is imperative that youth learn the history of flight; through such study, they will grow into adults who appreciate the significance of aviation. As a result, they also will respect the need for general aviation airports, which are almost an endangered species. Moreover, the students' studies of flight will propel many of them to choose aerospace careers as military pilots or engineers developing new aviation technology. Our country will benefit from this knowledge, which will result in the production and manning of aircraft vital to protection of the United States. Similarly, aerospace education teaches students about the technology behind satellites; without this knowledge, students will grow up ignorant of both the capabilities of space technology as well as the threat of its misuse by hostile countries. As you can see, aerospace education is extremely important to national security.*

Your next paragraph will begin with a topic sentence that highlights aerospace education's impact on your second point, space exploration, and so on.

THE OPENING PARAGRAPH

OBJECTIVES:

10. Defend the importance of having a strong beginning to your essay or speech.
11. Define "thesis statement."

If your audience is your teacher at school or your squadron commander, they don't have a choice but to read your work. But out in the "real world," your audience does have a choice, and if the first sentences of your essay don't pique their interest, they'll give up. Therefore, **the first few sentences should really grab the reader with an unexpected or dramatic scenario or story.**

For example, in an essay on children who live amidst filth, you might start with a few sentences like these:

> *Have you ever seen movies about children who live in filth and pollution? They walk through sewage and broken bottles to get to school. Clean drinking water eludes these youths, and life becomes a day-to-day struggle just to survive.* ***For me, this is no movie; this is everyday life ...***

Suddenly, the audience discovers the essay is not about a movie that might be fictional; it's about you! Your reader will very likely keep reading because now your essay has a personal touch.

Or perhaps your topic is leadership. Let's consider two ways you could begin your essay. Which is better?

> *Leadership is really important. There are three different aspects of leadership. The aspects of leadership are motivation, responsibility and dedication.*

Or:

> *Despite enemy machine guns ready to fire upon them, the soldiers ran from the trench and into certain death. Blood was everywhere; bodies lay all over the place. Then the messenger arrived at the front with news: the commander decided to delay the attack. Tragically, it was a message that arrived too late. Poor leadership can cause unnecessary death and destruction. And even when deaths don't occur, ineffective leadership can still harm the team.*

Obviously, the second opening will catch a reader's attention more than the first. It's more dramatic and explains why the topic is important and deserving of attention.

GRAMMAR ERRORS

Your use of the English language reflects the accuracy and credibility of your writing. Be careful to avoid these major grammar errors:

Comma splice: Cadet Simpson picked up the box of Meals Ready to Eat, she helped prepare meals for several cadets.

Problem and solution: The comma joins two independent clauses that could easily stand alone as separate sentences. To remedy, make each clause a separate sentence or use a conjunction such as "and" after the comma.

Run-on sentence: The search-and-rescue (SAR) lasted several days all the cadets were exhausted by the end of the week.

Problem and solution: The first independent clause ended with the word "days," but then the sentence "runs on" without punctuation. The error is easily remedied by placing a period after "days" and starting a new sentence with the word "all," which would be capitalized.

Pronoun-antecedent disagreement: A cadet must attend an encampment in order to secure their Mitchell Award.

Problem and solution: The pronoun "Their" refers back to "cadet," but the two words disagree in number. "Cadet" is singular and "their" is plural. The sticky issue is changing the "their" to "he" or "she" without upsetting either the males or females. To resolve this issue, make "cadet" plural (cadets) so that it can refer to male and female cadets.

THE THESIS STATMENT

So assume you've locked in the reader's attention after an excellent start to your essay. You still need to make a point. If you recall, your essay is often written to inform (teach) or persuade (to attempt to change someone's mind about an issue). To this end, your opening paragraph also should contain a **thesis statement**. *The thesis statement "is the central message of an essay" and your essay's "main idea."*[12]

For instance, in the sample outline and essay (pages 164-165), the thesis reads: *As Americans, we can see the benefits of flight perhaps more clearly than any other people in the world.*

When your readers see the thesis, they may have an "Aha!" moment, realizing exactly what's coming. For instance, in regard again to the sample essay, the title gives readers a strong indication they will be reading about flight; now, after reading your thesis, they are certain that you will be persuading, them of the *benefits of flight*.

The thesis statement can be more detailed as well, giving readers an even clearer picture of the remainder of your essay. It can contain **a blueprint, which is a "list of the ideas in your topic sentences."**[13]

Like a Playwright
Starting your essay with some drama will energize and encourage your audience.

Let's say you're writing an essay about your preference for glider orientation flights over powered flights. Thus, you have narrowed your topic to the advantages of glider orientation flights, which is your main idea. That being the case, your thesis might look like this:

> *While powered and glider orientation flights are both beneficial, glider flights far exceed the powered aircraft sorties as training opportunities because* **there is more time for one-on-one discussion between pilot and student,** *the student can learn basic flight in a more relaxed atmosphere,* **and there is an even greater opportunity to witness beautiful scenery**.

> "A good thesis produces an Aha! moment."

This type of thesis statement is ideal. It introduces the main idea of your essay, and it introduces what will become the three topic sentences that begin your body paragraphs.

For instance, the first sentence of your second paragraph (your first body paragraph), could say "Since gliders travel more slowly than powered aircraft, there are increased opportunities for discussion between pilot and student."

That is not to say that a glider is a lot slower than a Cessna or that no fruitful discussion occurs in a powered aircraft, but the atmosphere in the glider may be slightly more conducive to interaction.

TRANSITIONS IN BODY PARAGRAPHS

OBJECTIVE:
12.   Define "transitional words" and list five examples.

Once you've launched your opening paragraph, the body of your essay will follow easily. But since it will have at least three segments (paragraphs), each with a slightly different topic sentence, *transitions will be helpful in connecting your main points so that your essay flows easily from one point to the next.*

Transitions tell the audience when you have finished relating one idea and are switching to a new thought.[14] They are phrases and words like "In addition to," "However," and "Therefore," which also indicate you are using logic or thinking in your presentation. "Therefore," for instance, tells the audience that you have presented several ideas that, taken together, lead to a certain conclusion.

Transitions perform double duty, working within your paragraphs and between paragraphs.

For instance, as the sample aerospace essay moves from the topic of U.S. military airpower contributions to the importance of civilian aircraft, it might include this sentence: "Now that we have illustrated the U.S. military's contributions to airpower, let's look at civilian aviation's assets."[15] In fact, the transition would work well in a speech, too.

Take another look at our paragraph on the importance of aerospace education and note the transition words in bold:

> *Aerospace education is vital to the defense of our country.* ***First****, it is imperative that youngsters learn the history of flight; through such study, they will grow into adults who appreciate the significance of aviation.* ***As a result****, they also will respect the need for general aviation airports, which are almost an endangered species.* ***Moreover****, the students' studies of flight will propel many of them to choose aerospace careers as military pilots or engineers developing new aviation technology. Our country will*

AND NOW TO CONSIDER TRANSITIONS

Transitions connect your main points so that your writing flows easily from one point to the next. Some examples:

In addition to...

In addition to drill, cadets also learn about aviation...

However...

However, not all CAP pilots come from a military background. Some, like Capt John Curry, hail from the airlines...

Therefore...

Therefore, it's best to show respect to everyone you meet, regardless of appearances. You never know when you're in the company of heroes...

THE MIRACLE OF FLIGHT

Imagine traveling to visit your relatives across the nation. It's a 2,000 mile trip, and you've got to ride in your rusty red Model T. It's the early 1900s and some brothers named Wright are working on a flying machine. Maybe one day you can use that invention to enjoy an easier trip from New York to Oregon. For now, you can only foresee a bumpy ride and extremes of hot and cold and hills and lonely flatlands. Yes, luckily this is just a bad dream. Flight has transformed the lives of billions of people and the nations they call home. **As Americans, we can see the benefits of flight more clearly than any other people in the world.** In the United States, the advent of flight has eased the burden of travel, strengthened our military and promoted general aviation.

Thanks to the invention of large aircraft capable of carrying dozens to hundreds of people, travel has never been more convenient. Businesspeople in Waco, Texas, can fly to Tulsa, Oklahoma, in a single morning, trades and deals can be made and these people can be back home with their families in the evening. Meanwhile, families who have been separated because of military service and new employment opportunities may now live on opposite sides of the country. But although it may be expensive, they can board aircraft and fly to see each other within a day. Lands once only accessible by boat can be reached easily by plane. Students and adults can fly to places like London, Italy or Japan in less than a day. As a result, Americans can know their own country's history intimately as well as the story of other lands' growth and development. **Flight, then, has made conducting business, enjoying family bonds and touring the country and world easier.**

At the same time, flight has boosted American military effectiveness as well. What if the Wright Brothers had worked solely with bicycles and not cared about flight? What if others, like Charles Lindbergh and Amelia Earhart, had shown no interest in flight? And what if other countries capitalized on flight first and focused especially on creating war machines that could attack the United States? Fortunately, the above-mentioned aviators and thousands of others took to the air in the early 1900s and ensured that Americans would eventually establish air supremacy. Controlling airspace above World War I and II battlefields and oceans helped the United States enjoy victories in both wars. In the modern day, military jets can be dispatched across the world instantly to combat and deter enemy forces. Drones can be launched from Nevada to Afghanistan in just minutes to attack terrorists. Rockets can be used to shoot down incoming missiles. **There is no doubt that aircraft have made America a much safer place to live.** Yet aircraft are not only an often violent means of defense; they also are a tool for leisure.

Thanks to the advent of aircraft, general aviation in the United States has thrived since the first hometown airports began to appear on maps. Choose any day of the week, but especially a Saturday or Sunday, go to any general aviation (GA) airport, and you will find young people learning to fly and often see older aviators enjoying conversation or washing their airplanes. Not only does general aviation allow for leisure, but in a small Piper or Cessna, one can fly relatively quickly and inexpensively to neighboring states to vacation or visit relatives. In addition, the presence of general aviation heightens interest in flight and aerospace for youth because it is accessible; you don't have to seek a military career to enjoy flight. In countries that don't have general aviation, military careers are the only route to enjoying flight. **By contrast, the advent of flight has propelled general aviation in America, where GA airports add to the variety of our spice of life.**

Of course despite all of those who love flight, there's still a handful who absolutely despise the invention. For instance, some critics believe that the flying machine has made war more violent. They cite the dropping of the atomic bomb on Japan, for instance. In reality, one can also argue that flight has made war more humane. Trench warfare for centuries cost the lives of soldiers around the world; infantry living in trenches for months faced disease and almost certain death. The modern soldier can rely less on the trench because superior airpower greatly limits the need for hand-to-hand combat typical in trench warfare. So, on the whole, flight has improved, not worsened, the impact of wars.

As you can see, Americans have truly capitalized on the advent of flight in the most remarkable ways. Flight has brought Americans closer together; in the matter of a morning, businesspeople can cross state lines in a business jet or airliner to cut a deal, and relatives can travel hundreds of miles to share an embrace. We can sleep safer at night because the flight of military aircraft makes us more secure. When day arrives, and the sun shines on the wings of little planes and the town airport, the general aviation community starts another day of training new pilots and providing a haven where aviators can share a war story, the latest news and a hot dog or hamburger. Thanks to pioneers like the Wright Brothers, Ms. Earhart, and Mr. Lindbergh, the stage was set for space exploration and the many, many more feats that will accomplished in aviation in the 21st century.

*benefit from this knowledge, which will result in the pro
duction and manning of aircraft vital to protection of the
United States.* **Similarly***, aerospace education teaches
students about the technology behind satellites; without
this knowledge, students will grow up ignorant of both the
capabilities of space technology and the threat of its misuse
by hostile countries.* **As you can see***, aerospace education
is extremely important to the security of our nation.*

Transitions within the paragraph connect thoughts of individual sentences. For instance, shortly after the paragraph begins, the essay relates that young people will grow to appreciate aviation. Then, the following sentences states "as a result," youth "will respect the need for general aviation airports." The transition, "as a result," becomes a bridge between two sentences that shows a cause-and-effect relationship; in other words, the young people's *appreciation* for general aviation *will cause them to respect* the existence of general aviation airports.

Transitions do act like bridges between sentences; but just as important, they link paragraphs together.

Take the above paragraph for example. This paragraph being essentially complete, the next paragraph is supposed to relate that aerospace education nurtures space exploration.

If we go straight from speaking of national defense to the topic of space exploration, the pleasant flow of the essay will be interrupted. But if we amend the last sentence of the paragraph just shown, we can create another bridge that carries the reader's mind from thoughts of national defense at the end of one paragraph to considerations regarding space exploration beginning the next paragraph.

> **As you can see***, aerospace education is
> extremely important to the security of our nation; of
> equal importance, however, is aerospace education's
> impact on space exploration.* **(end of paragraph)**
> *Teaching students about space travel and
> exploration broadens their understanding of flight and
> motivates their study of the universe.* **(beginning of
> next paragraph)**

See, it's easy! You ended one paragraph and simultaneously introduced the topic of the following paragraph. Not so fast, though. Yes, by now you've presented three logical arguments to support your thesis, but be careful not to assume the reader will agree with every point you make.

SAMPLE OUTLINE

Topic: The miracle of flight

Thesis: As Americans, we can see the benefits of flight perhaps more clearly than any other people in the world.

I. Opening paragraph and thesis

II. Travel is more convenient
(1st body paragraph)
   A. Businesspeople benefit
   B. Families stay in touch

III. Military air supremacy
(2nd body paragraph)
   A. Improves national security

IV. General aviation flourishes
(3rd body paragraph)
   A. Youths learn to fly
   B. State-to-state travel occurs

V. Objection
(4th body paragraph)
   A. Critics say flight increases
      violence of war
   B. Actually war is more
      humane now
      (rebuttal????)

VI. Conclusion

Transitions
Like a bridge connecting one side of the bay to another, transitional statements connect paragraphs.

ANTICPATING & REFUTING OBJECTIONS

OBJECTIVE:
13. Describe what an objection is and explain its importance.

Anyone who has watched court dramas on television knows the word "objection." The attorney tells the jurors, "The defendant lives in filth and represents a moral stain upon our community." But the opposing attorney suddenly cries, "Objection, your honor! How my client lives and acts has nothing to do with whether or not he knows how to fly the airplane that collided with the train!"

Anticipating objections, which are reasons or arguments presented in opposition, improve your communication as well.[16] If you don't expect objections or disagreements, you are fooling yourself. On the other hand, understanding that not everyone will agree with your writing (or speech for that matter) demonstrates your maturity. Hence, you may work really hard to advocate the continued existence of general aviation airports; however, by recognizing that some neighbors of airports despise them, you show your understanding of the complexity surrounding the issue of their presence among homes.

To balance your essay between your arguments and the potential disagreements you may face, you can anticipate the most likely point of disagreement and use the opportunity to provide additional support for your thesis.

In the next-to-last paragraph of the sample essay (page 164), a typical objection to the benefits of aviation is cited. That is, some people feel advances in aviation have led to greater death and destruction in war. The objection, if supported adequately, has merit, but the essay refutes the concern by emphasizing the carnage caused by trench warfare.

> *Trench warfare for centuries cost the lives of soldiers around the world; infantry living in trenches for months or years faced disease and almost certain death. The modern soldier can rely less on the trench because superior airpower greatly limits the need for hand-to-hand combat typical in trench warfare.*

In the end, essayists show respect for others – even those who may oppose them — by admitting that their thesis may not be flawless, but they can still use the objection to further advance their own cause. Once again, CAP's Core Values – in this instance, Excellence and Respect – play an important role in communication.

THE CONCLUSION

OBJECTIVE:

14. Describe the function of the conclusion.

You've pleased your audience with the support for your thesis. Now you quit writing, right? Not quite. That would be like stopping mid-sentence in a conversation. **In your conclusion, you reiterate your thesis but use different words.** Take the opportunity as well to restate your main points (your topic sentences) and ***think of something unique and memorable with which to end your essay.***

Look once more at the sample essay (page 164). The last paragraph begins with several sentences that are clear summaries of the main points in the body. Then, the essay concludes with the realization that aviation pioneers set the stage for all other future accomplishments in aviation.

HEADING OFF OBJECTIONS

State your points and people will naturally be swayed to your way of thinking, right? Hardly. If your audience is familiar with your topic they may already have objections to your ideas. Wise leaders anticipate those objections and head them off.

By taking the initiative to identify possible objections and then arguing why they are not fatal flaws, you show a full command of the issue's complexities. Head off objections to become a more effective advocate.

Some examples of dealing with objections:

THE UNIFORM
[Objection] The uniform is the mark of the military. It's the attire of warfighters, not youth. Cadets could learn teamwork and discipline even if outfitted in jeans and a polo shirt. *[Counter-argument]* These criticisms make sense at first glance, but they overlook the uniform's unique power to motivate young people to rise above the ordinary. There's nothing inspiring about jeans and a polo.

COLLEGE CHOICE
[Objection] Today, even average students are expected to attend a 4-year college. The conventional wisdom is that if you don't attend a traditional college straight out of high school you'll be left behind. *[Counter-argument]* But junior colleges and technical schools afford students a better opportunity to figure out what they really want to do before investing a huge sum in a big, impersonal university. Moreover, the credits can be transferred if the student later decides that the traditional 4-year college is right for them.

CADETS & ES MISSIONS
[Objection] CAP's emergency services missions are often matters of life and death. They're no place for children, whose mere presence is a safety hazard. Why then does CAP allow cadets to participate in missions? *[Counter-argument]* Individuals mature at different rates – some adults don't measure up to the professionalism of our top cadets. Moreover, if you were lost in the woods, wouldn't you want every available person who has been trained for the job to be searching for you?

TIPS FOR WRITERS

- Don't use no double negatives.

- Make each pronoun agree with their antecedent.

- Joined clauses correctly, like a conjunction should.

- About them sentence fragments.

- When dangling, watch your participles.

- Verbs has to agree with their subjects.

- Just between you and I, case is important.

- Don't write run-on sentences they are hard to read.

- Don't use commas, which aren't necessary.

- Try not to ever split infinitives.

- Correct speling is essential.

- Proofread your writing to see if any words out.

- A preposition is a poor word to end a sentence with.

- Don't be redundundant.

— Anonymous

OBJECTIVES:

15.   Explain the purpose of a staff study.

16.   Identify the sections required in a staff study and discuss the purpose of each.

As a leader, you might use writing to recommend changes or improvements. The temptation may be to use email to vent concerns. Instead, if you have gripes or suggestions, there are great ways outside of email to make them known. For instance, members of the Air Force share their concerns in several professional formats, one of them being the staff study. You can do the same in Civil Air Patrol.

STAFF STUDY

Consider this scenario. You enjoy CAP; you've been a member for two years in your local squadron. But the meetings are tedious. There's too much drill and not enough "fun" activities, such as paper airplane contests and rocket launches. You could say nothing, and, as a result, get increasingly frustrated until one day you yell at the squadron commander. Hopefully you won't, but there might be a temptation for you to vent in an email to one of your cadet peers, using harsh words about the commander. But then one day your peer accidentally forwards your anger-filled email to the squadron commander. He's furious. Not only have you not helped your cause, you may have permanently injured your reputation.

Purpose of the Staff Study Report. ***The staff study provides a professional format for presenting concerns and solutions.*** It's a tool for leaders to use when they want to talk about a problem and offer a solution thoroughly and in a logical manner. When you offer solutions to a problem, your superiors will view you with greater respect. They will see that you care; they will recognize that you are not whining but trying to improve the squadron.

Difference Maker
You can be the force behind positive change through effective writing.

7 STEPS *for* EFFECTIVE COMMUNICATION

★ Analyze purpose & audience

★ Research your topic

★ Support your ideas

★ Organize & outline

★ Draft

★ Edit

★ Fight for feedback & win approval

PRINCIPLES IN WRITING STAFF STUDY REPORTS

Consider Your Audience. In most cases, you're probably addressing your squadron commander. What concerns is he or she likely to bring to your problem?

State the Problem. Identify your problem clearly. If you can't summarize it in one or two sentences, try again. Also, be very precise in your problem statement.

EXAMPLE: Cadet attendance is down 40% compared with the attendance level six months ago.

Analyze the Entire Problem. What factors drive cadet attendance at weekly meetings? Transportation? The meeting time? Meeting location? The activities offered? How well the meeting is planned? Consider these issues as you develop the facts, assumptions, criteria, and definiton sections.

Gather Data. Gather any information that is related to the problem. For example, you might survey cadets and report on the state of their morale. You could compare some recent meeting schedules and activities offered with the schedules and activities when attendance was high six months ago.

Identify the Facts. Remember that facts must be provable –they aren't opinions.

EXAMPLE: When attendance was at its peak, the squadron was engaged in the AEX program and color guard training; today we are not working any major projects.

EXAMPLE: Survey data shows that 8 out of 10 inactive cadets believe the meetings have become boring and lack hands-on activities.

Identify Assumptions. An assumption is something that relates to the problem and most people accept it as true, even if it can't be proven concretely.

> **"Facts must be provable – they aren't opinions."**

CIVIL AIR PATROL APEX CADET SQUADRON
UNITED STATES AIR FORCE AUXILIARY
821 Buck Jones Road
Raleigh, NC 27606

[DATE]

MEMORANDUM FOR [OFFICE SYMBOL OF ADDRESSEE]

FROM: [OFFICE SYMBOL OF AUTHOR]

SUBJECT: The Staff Study Report

PROBLEM

1. Clearly and concisely state the problem you are trying to solve.

FACTORS BEARING ON THE PROBLEM

2. **Facts.** Identify verifiable facts that directly relate to the problem.

3. **Assumptions.** Identify those facets of the problem that most knowledgeable people accept as true, although they can't be firmly proven.

4. **Criteria.** Identify the standards, regulatory requirements, operational needs, or limitations you will use to test possible solutions.

5. **Definitions.** Explain technical terms that may confuse your audience.

DISCUSSION

6. This is the meat of your argument. Offer your perspective on the problem and explain your solution or possible solutions. Show the logic of your thinking.

CONCLUSION

7. State your conclusion as a workable, complete solution to the problem you identified in paragraph 1. The conclusion statement is especially important if you discussed the pros and cons of several possible solutions in the discussion section.

ACTION RECOMMENDED

8. Tell the reader the action(s) necessary to implement the soltuion. Make it easy for the boss; this section should be clear enough so that the boss can simply write, "Go!" or "Approved" and all stakeholders will know their assignments in implementing the solution.

John F. Curry
JOHN F. CURRY, C/CMSgt, CAP
Cadet First Sergeant

Atttachments (if needed)

EXAMPLE: While recruiting more cadets will boost the attendance figures in the short term, it won't address the underlying problem of why current cadets aren't attending metings.

Identify the Criteria for the Solution. Include standards, requirements or limitations you will use to test possible solutions.

EXAMPLE: In planning weekly meetings, the *Cadet Staff Handbook* advises units to develop a written schedule at least one week in advance. This schedule needs to be coordinated among all stakeholders and approved by the commander.

EXAMPLE: To avoid boredom and to ensure quality training, drill should be limited to 15 minutes per meeting, as sugested by the *Cadet Drill Guide*.

Define Key Terms. If using any technical terms or jargon that your audience may not be familiar with, define them.

EXAMPLE: A "hands-on" activity is one where cadets are actively engaged in doing something, such as building a rocket, solving a team leadership problem, or debating a Core Values issue. Hands-on activities are not passive and boring, like lectures or endless PowerPoint presentations.

List Possible Solutions. There may be several ways to approach your problem. A good leader will consider all the options. In the discussion section, consider the pros and cons of a small handful of possible solutions.

EXAMPLE: (1) Adopt a multi-week, hands-on project to provide structure and purpose to the meetings... (2) Authorize the cadet commander to draft and coordinate a written meeting schedule one week in advance... (3) Use multi-voting to survey cadets' interests and set activity goals for the coming quarter...

Test Possible Solutions. In the discussion or conclusion section, explain how each possible solution measures up against your criteria.

Recommend Action. Close by recommending a specific course of action. Make it clear who is to do what and write in such a way that the boss can simply reply, "Great idea. Go for it!"

EXAMPLE: The squadron commander should endorse the attached plan for establishing a color guard. Specifically, (1) C/2d Lt Curry, working under Capt Arnold, will lead the effort; (2) the squadron will make $300 available for color guard supplies; and (3) the cadet commander will invite all Phase I and II cadets to try-out for the color guard on 1 December.

The staff study takes time, but when you undertake the effort, it will set you apart as a leader who wants to improve the team, not an individual set on complaining about everything.

Jargon.
special terms that convey a very specific meaning within a profession or field of study

---

STEPS OF PROBLEM SOLVING

1. Recognize the problem
2. Gather data
3. List possible solutions
4. Test possible solutions
5. Select final solution
6. Act

BODY OF STAFF STUDY

1. Problem
2. Factors bearing on the problem
3. Discussion
4. Conclusion (a brief restatement of final solution)
5. Action recommended

EMAIL & PROFESSIONALISM

OBJECTIVES:

17. Discuss the pros and cons of email as a communication medium.
18. Describe guidelines for maintaining professionalism when using email.

Often something so formal isn't necessary. In many instances, you can communicate concerns of a lesser nature via email.

Benefits of Email. In today's workplace, email is arguably the most popular means of communication. In organizations like CAP where the members are geographically separated, email is even more indespensible. Some benefits of email include:

★ Email is essentially free, assuming you have access to a computer and the Web

★ Unlike the phone, email is unobtrusive; you can email someone at 2am and not worry about waking them up

★ As a form of written communication, email is easy to file and refer back to again later

★ Email is easy to share; newcomers to a project can review previous email traffic and catch up on what's going on

Pitfalls of Email. At the same time, email is not without its pitfalls. Practically every modern leader has lived to regret a mistake made in their email. One of the best things a leader can do to safeguard their reputation for professionalism is to be mindful of the pitfalls of email, including these:

★ Because email is so quick, people tend not to scrutinize their writing and therefore they don't say precisely what they mean.

★ It's easy to hit "send" instead of rethinking what you've written. This is especially true for controversial or emotional issues.

★ Email can be a cop-out; instead of working out personal differences face-to-face, people may hide behind email.

★ Email tends to snowball, especially as multiple individuals are included in the distribution; instead of being a helpful tool, the inbox quickly fills up and email becomes a chore.

★ You can't be certain that when the recipient will see your message, or even at all. The telephone remains the most effective means for dealing with time-critical issues when people are in different places.

GUIDELINES *for* PROFESSIONALISM *in* EMAIL

★ Be brief. Email is best suited for quick, short messages.

★ Address superiors as sir, ma'am, or by grade, just like you would in person.

★ Use proper spelling and grammar. Do not use emoticons. Also, avoid weird fonts and colors.

★ Reply to emails promptly, within 48 hours if at all possible.

★ If you have a question, ask it directly. If asked a question, answer it directly.

★ Be judicious about sending copies. Rather than using "reply all," if a conversation affects only two or three people, reply only to them, not the whole group.

★ Don't get into arguments, tirades, or make unprofessional remarks. If confronted with a sensitive or emotional issue, have the courage to talk with the other person face to face.

★ Don't forward jokes, spam, rumors, etc., unless the other person is a good friend who welcomes such messages.

★ Close with a signature block that includes your name, title, and telephone number.

★ Think before you type and think before you push "send." Once the email is out in the world, there's no getting it back or controlling who else sees it.

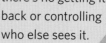

MAKING EMAIL WORK FOR YOU

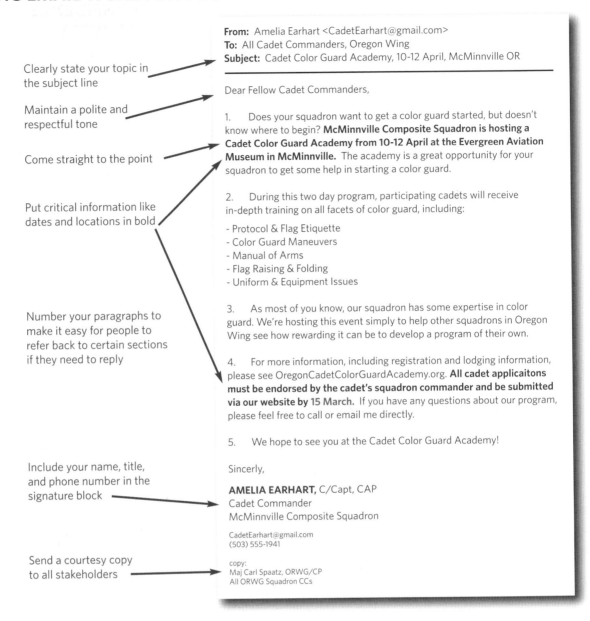

Clearly state your topic in the subject line

Maintain a polite and respectful tone

Come straight to the point

Put critical information like dates and locations in bold

Number your paragraphs to make it easy for people to refer back to certain sections if they need to reply

Include your name, title, and phone number in the signature block

Send a courtesy copy to all stakeholders

From: Amelia Earhart <CadetEarhart@gmail.com>
To: All Cadet Commanders, Oregon Wing
Subject: Cadet Color Guard Academy, 10-12 April, McMinnville OR

Dear Fellow Cadet Commanders,

1. Does your squadron want to get a color guard started, but doesn't know where to begin? **McMinnville Composite Squadron is hosting a Cadet Color Guard Academy from 10-12 April at the Evergreen Aviation Museum in McMinnville.** The academy is a great opportunity for your squadron to get some help in starting a color guard.

2. During this two day program, participating cadets will receive in-depth training on all facets of color guard, including:

- Protocol & Flag Etiquette
- Color Guard Maneuvers
- Manual of Arms
- Flag Raising & Folding
- Uniform & Equipment Issues

3. As most of you know, our squadron has some expertise in color guard. We're hosting this event simply to help other squadrons in Oregon Wing see how rewarding it can be to develop a program of their own.

4. For more information, including registration and lodging information, please see OregonCadetColorGuardAcademy.org. **All cadet applicaitons must be endorsed by the cadet's squadron commander and be submitted via our website by 15 March.** If you have any questions about our program, please feel free to call or email me directly.

5. We hope to see you at the Cadet Color Guard Academy!

Sincerly,

AMELIA EARHART, C/Capt, CAP
Cadet Commander
McMinnville Composite Squadron

CadetEarhart@gmail.com
(503) 555-1941

copy:
Maj Carl Spaatz, ORWG/CP
All ORWG Squadron CCs

PUBLIC SPEAKING

OBJECTIVE:

19. Describe ways to combat stage fright.

If you are speaking for the first time in front of a group, you may get s*tage fright – an anxiety of speaking in front of people.*[17] Right before you are about to speak, suddenly your heart starts to race, you can't catch your breath, and perhaps you fear even entering the room where you will speak. It's okay, everyone has felt this way at one time or another.

Since likely the worst thing is to go up to a podium and not know what to say, it makes sense that b*y adequately preparing your speech, you can decrease your stage fright up to 75 percent.*[18] Let's consider how you can get ready for public speaking.

COMMON SPEAKING METHODS

OBJECTIVE:

20. Describe the four most common formats of a speech.

When speakers present a speech, they employ of one of four common methods: reading from a manuscript, speaking from memory, speaking without specific preparation, or speaking extemporaneously with preparation.

Manuscript. Of all of these, reading straight from a manuscript is the poorest method. *It is employed only when the material being conveyed is so important or complex that an inaccurate phrase might cause a great misunderstanding.* In this scenario, you benefit from saying exactly what you want, but at the expense of intimacy and flexibility.

Memory. Others choose to speak from memory, which also isn't usually a good idea. A memorized speech is difficult to deliver without sounding monotone and flat. *The speaker becomes overwhelmed with accurately stating the speech as it was memorized, so he loses spontaneity.* The memorization process iteself is also extremely time-consuming.

Impromptu. On the other hand, one of the best and most challenging communication venues is the impromptu speech, also called an elevator speech. *For this delivery, the speaker is given a topic and only a few minutes (or less!) to gather his thoughts before speaking.* The impromptu speech is what leaders face most often in everyday life. How many times have you been in uniform in public when a curious bystander approached you to ask about CAP? With impromptu speaking, leaders have to think on their feet.

Extemporaneous. The most common type of formal speech and the one that usually yields the best results is the extemporaneous speech. One carefully plans and outlines this speech using strategies suggested earlier in the chapter. *Extemporaneous speakers study their outline in depth, but instead of planning what they'll say word-for-word, they grant themselves freedom to be spontaneous.*

CONFIDENCE *in* DELIVERY

Nearly everyone who has stood before a crowd has felt nervous prior to speaking. However, good preparation can help as well as the other hints below:[19]

1. **Check** your equipment beforehand. Perhaps nothing peeves an audience more than waiting while you fix your projector or computer.

2. **Practice** repeatedly, especially in front of peers willing to "act" as an audience.

3. **Memorize** your introduction and your transition into your first point.

4. **Smile** and relax. Even if you are nervous, a smile can help hide your fear from the audience.

5. **Take** a brief walk before you begin your speech.

6. **Make** eye contact with audience members.

7. **Involve** the audience by asking questions and seeking their opinion of your topic ("How do you feel about general aviation airports?")

8. **Look** neat and tidy. It will boost your confidence.

OUTLINING A SPEECH

OBJECTIVE:
21. Identify and describe the parts of a speech.

Your organization begins with an outline that will differ just slightly from your essay outline.

The Specific Purpose. First, begin with *a specific purpose, a clear statement of what you hope to accomplish as a result of your speech.*[15] Again, like writing, there are three main purposes in public speaking: to entertain, to inform, or to persuade.

The Central Idea. In public speaking, *the central idea is like the thesis statement used in writing. It is a compact expression of your argument.* It's your main point, so you may want to state it more than once during the course of your talk to ensure the audience (who, unlike readers, can't go back and review your main point on their own) understands your message.

Got Edits?
The most outstanding speakers edit, revise, tweak, and edit their remarks some more.

Introduction. Many of the principles writers use to craft their introductions apply to speakers as well. However, it's especially important for speakers to include an overview in their introduction. During the overview, the speaker clearly identifies the subject of the talk and lists some of the main points that will be made. Once again, we see how speakers must be mindful that their audience cannot review or skim ahead, as readers can.

Body & Conclusion. As with written communications, a speech should include a body and conclusion. Because this chapter already covered those topics in the section on writing, we won't revisit those points here.

SIGNPOSTS

OBJECTIVE:
22. Describe the term "signpost," and explain its function.

To keep the audience engaged and to help them follow your argument, effective speakers use signposts – brief verbal cues indicating your progress through an outline.[20]

*"I've already told you about **two of CAP's main missions**, emergency services and aerospace education. **My third point** is that CAP's Cadet Program . . ."*

*"The basics of first aid come down to the ABCs. That's airway, breathing, and circulation. **First, let's talk about A for airway...**"*

Restatements. The restatement is a type of signpost that speakers use to emphasize their key points. If someone says something twice, you know that it must be important. Some examples:

*"**Cadets go on to lead incredible lives.** Former cadet Nicole Malachowski became the first female Thunderbird pilot. Former cadet Eric Boe piloted the Space Shuttle. Former cadet Ted Bowlds became a 3-star general in the Air Force. Every long-time senior member knows a similar tale. **Cadets go on to lead incredible lives.**"*

Signposts are more important in public speaking than they are in writing because readers can re-read confusing passages, use the margins to number the main points, use a highlighter to mark key passages, and the like. In contrast, an audience listening to the speaker lacks those benefits, so it is up to the speaker to help the audience follow along.

THE CONCLUSION

OBJECTIVE:
23. Discuss principles in concluding a speech.

After you have presented your evidence in the body of your speech, it is then time to summarize the points you made and end your speech. Again, the audience's needs come into play; speakers need to reiterate their main points to remind the audience because the audience cannot easily review those points on their own. To wrap up a persuasive or informative speech, a good technique is to leave the audience with something you want them to believe or to do.[21]

A fine conclusion to an informative talk:

> *As you have seen, the facts I've presented demonstrate the positive impact of aerospace power on national security. First, airpower allows us to use military force, as a last resort, while minimizing the loss of innocent lives. Second, because modern aircraft fly faster than sound, airpower allows us to strike on very short notice. And third, airpower's newest platforms, the unmanned aerial vehichles, allow us to minimize our own losses in a fight. But now that you know about airpower, what can you do? I encourage you to write your elected leaders and tell them you support the X-99 appropriations bill currently before Congress.*

In an casual speech or one whose purpose is merely to entertain, a good way to wrap up is to leave the audience with a zinger – a memorable or humorous last line.

STOP

✈ EXIT 159B

SIGNAL YOUR TRANSITIONS

EXIT ONLY ⬇

COMMUNICATING FOR YOUR CAREER & LIFE

Effective communication can enrich your career opportunities and therefore shape your life. Not only is communication a vital part of many occupations – presentations that must be made, reports to be written, and more – but before you ever land your dream job, you'll have to communicate through a résumé and interview.

THE RÉSUMÉ

OBJECTIVES:
24. Describe the purpose of a résumé.
25. Identify the major components of a résumé.

A résumé briefly documents your work history and gives you the opportunity to show what makes you qualified for a job. That's a lot for a short, one-page document to accomplish. Résumés require lots of editing and review. Let's consider the major parts of the résumé:

Personal Information. Always list your name, address, phone number, and email address at the top of the résumé. However, ***it is not appropriate to list additional personal information like your age, height and weight, marital status, etc.***

Objective. ***Some résumé experts suggest you identify your career aspirations and/or immediate objective.*** For young adults, these sections can help compensate for your relative lack of experience because it shows the hiring manager that you have clear goals and interests.

Education. As a student, ***list your highest level of education.*** Once you enter the adult workforce, list all college degrees and professional certifications. Again, young adults who need to compensate for their inexperience may want to amplify their educational credentials by including their grade point average and class rank.

Experience. Most hiring managers consider this section to be the meat of the résumé. There are two primary ways to complete the experience section. First is ***the career chronology method in which you list the various positions you've held and outline the major accomplishments of each.*** The career chronology method is the most common résumé style in the adult workforce. Second is ***the skills inventory method in which you focus on the work-related skills and abilities you've acquired.*** The skills inventory approach is rarely used by adults, but some experts suggest it to students who possess only a modest degree of career experience.

Your résumé represents you. Carefully craft your résumé – it's likely to be the determining factor in whether you receive a job interview.

Awards, Honors & Extra-Curricular Activities. This is another section whose content and relative importance will vary depending on whether the applicant is a student or an adult with an established career. ***Students should list all awards and honors that are relative to the position being sought.*** For example, a student applying for a summer job at the airport should mention their CAP experience, but their experience on the chess team can be omitted.

References. It is customary to list two or three individuals who can speak to your professionalism, work-related skills, or employment history. However, ***before using someone as a reference, be sure to ask their permission and verify that indeed they'll recommend you for the position.***

If your résumé impresses your potential employer, then you can work on the next step: preparing for a job interview.

JOB INTERVIEWS

OBJECTIVES:
26. Explain the purpose of a job interview.
27. Describe ways applicants can prepare for interviews.
28. Discuss principles of etiquette for interviews.

The company has "seen" you on paper. Now the employer wants to meet you face-to-face. If you have made it this far, you have a good chance of landing the job. Next, let's consider some principles to guide you during a job interview.

Make a Good First Impression: ***Make yourself neat and presentable and dress in attire you would wear if selected for the position.*** The rule of thumb used to be to wear formal business attire to all interviews, but most career coaches today caution applicants not to overdress. Further, to show you are well-prepared for business, always bring extra copies of your résumé, plus a notepad to take notes about your discussion.

Prepare for the Classic Questions. It's no secret what questions hiring managers are apt to ask interviewees, so be ready for them. Some classic questions include:

Why should I hire you?
Tell me about yourself.
What experience do you have to prepare you for this job?
What do you know about our organization?
Where do you see yourself in five years?

JOB INTERVIEW ETIQUETTE

Professionalism is not the job you do but how you do it. That was this volume's opening line, and it's a truism that applies to interviews. Some tips on interview etiquette:

★ Be on time, or better yet, 10 minutes early

★ Check your coat, purse, and non-essential belongings

★ Turn off your cell phone

★ Don't bring food, drink, or gum to the interview

★ Shake hands firmly and look others in the eye

★ Give the interviewer your complete attention

★ Talk about the job and your qualifications first, and save discussion about pay for the end or the second interview

★ Don't gripe about your old boss

★ Be an adult - your parents have no place in job interviews, negotiations, or your relationship with the boss

★ Follow up with a thank you note

The interviewer(s) will use questions to try to get to know you better as a person. *They are gauging your communication skills, people skills, self-confidence, and other leadership traits as much as they are discerning whether you're a good match for the job's challenges.* Remember, the interview is not a friendly chit-chat, though it may be cordial. Therefore, try to answer the interviewer's questions in such a way as to highlight your strengths and qualifications.

Ask Good Questions. Too many applicants forget that an interview can be a two-way street. That is, *the applicant is interviewing the organization and the boss to see if they match the individual's needs and desires.* Therefore, come prepared to ask questions of your own. Some classic examples include:

What's the team's culture like, or the boss's leadership style?

What's the job really like on a day-to-day basis?

Is this a new position, or would I be replacing someone?

Is there room for growth in this positon?

SAMPLE RÉSUMÉ

An FBO at the local airport is looking for a part-time lineman, gofer, and receptionist. Cadet John Curry prepares his résumé for the job.

Includes name & contact info

The career goal and immediate objective show why Curry is interested in a job at the local airport

JOHN CURRY
79 Fenton Ave Laconia, NH 03246
(603) 555-1934 cell
johncurry@isp.com

CAREER GOAL
Aspiring air traffic controller or meteorologist

IMMEDIATE OBJECTIVE
Obtain a summer job that allows me to learn about aviation while saving for college

EDUCATION
Laconia High School, junior, 3.8 GPA, (top 10% of class)
Coursework includes triginometry, chemistry, & introduction to management.

WORK EXPERIENCE *[on a real resume, you'd include dates of service]*
Everett's Yard Services, landscaper (part-time, 24 hrs per week)
Performed yardwork and landscaping services for residential clients
Worked independently, often with no direct supervision at the jobsite
Completed safety training and qualified to operate potentially dangerous power tools & equipment

EXTRA-CURRICULAR ACTIVITIES

Civil Air Patrol, cadet second lieutenant (top 15% nation-wide)
Successfully completed introductory curriculum in aviation and team leadership
Served as a flight commander, responsible for the training, discipline, and welfare of fourteen
        cadets during a 7-day annual encampment
Participated in three flights in a Cessna 172 and one KC-135 aeriel refueling mission

Laconia High School Chess Team: co-captain
Co-organizer for the first annual Lakes Region Chess Academy, a one-day program that introduced
        30+ kids aged 9 to 13 to the fundamentals of chess

Parish Council Youth Representative, St. Joseph Catholic Church
Advise the pastor and adult council members on how to better serve teen parishioners
Served as mentor & role model for 8th grade Confirmation class

Cycling
Participated in dozens of bicycle road races; average 70 miles' training per week

AWARDS & HONORS
National Honor Society, Laconia High School
General Billy Mitchell Award, Civil Air Patrol
Honor Cadet, Civil Air Patrol Encampment, Pease International Tradeport

REFERENCES
Everett Lord, owner, Everett's Yard Service, 555-7981, everett@isp.com
Major Ira Eaker, Civil Air Patrol squadron commander, 555-1941, eaker@isp.com

SOCIAL MEDIA

OBJECTIVE:

29. Discuss pros and cons of social media's impact upon society.

As nearly every American teen knows, ***social media refers to online tools that allow users to publish and share content such as text, photography, and video over the Web.*** Twitter, Facebook, LinkedIn, and YouTube are just a few of the most popular sites. Blogs (web logs) and forums are other kinds of social media where readers and writers are encouraged to interact.

Benefits of Social Media. Although American society is highly mobile in the 21st century, people are able to keep in touch with friends and family around the nation or the world, thanks to social media. Such sites serve as ideal places for communication. People can instantly share ideas, advice, photos, and videos. Moreover, businesses and clubs use social media to promote their latest offerings. Many adults especially use sites like LinkedIn to advertise their talents and experience, hoping their profile might attract a potential employer or client. Social media offers job-seekers a new way to circumvent traditional hiring processes. And for extroverts — people who enjoy being in the midst of a crowd of people — social media enables them to become "friends" with thousands of people they've never met in person.

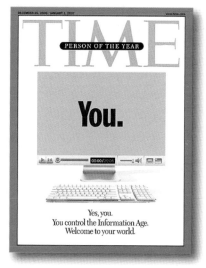

Person of the Year.
So revolutionary is social media that in 2006, *Time* magazine named "you" as its Person of the Year.

Drawbacks of Social Media. Do you really want a profit-seeking company to have access to your personal information? Privacy rights advocates warn that one consequence of social media is that it turns the previously private into public information. Moreover, data that is published online tends to remain online, even if "deleted." Many teens have learned the hard way that college recruiters, employers, and others use social media to learn more about young people; post photos of yourself doing something stupid and you might later regret it. Other social critics worry that because the Web offers anonymity – as a New Yorker cartoon famously quipped, "On the Internet, no one knows you're a dog," – normal standards of politeness and civil behavior are eroding. In online "flame wars," people impolitely disparage one another's remarks in web forums.

CONCLUSION

You can succeed as a communicator every day. After your essay or speech is turned in or delivered, it may fade away like the contrails of a jet. But every single day, you can speak confidently and maturely. You can make yourself heard in writing as well. Your success as a communicator will help you in CAP as you strive for promotions, as you tackle high school and college courses, and as you communicate in all aspects of life.

DRILL & CEREMONIES TRAINING REQUIREMENTS

As part of your study of this chapter, you will be tested on the principles and processes of the wing formation and review. Ask an experienced cadet to assist you in learning about this subject. For details, see the *USAF Drill and Ceremonies Manual*, available at capmembers.com/drill.

From the *Air Force Drill & Ceremonies Manual*, Chapter 6, Section C
Wing Formation and Review

ENDNOTES

1. Robert Evans, ed. Close Readings (Montgomery, AL: Court Street Press, 2001), 32-33.

2. President Gerald R. Ford, Remarks to the Veterans of Foreign Wars Annual Convention, Chicago, Illinois, August 1974 <http://www.presidency.ucsb.edu/ws/index.php?pid=4476>

3. Gerald R. Ford, 93, Dies; Led in Watergate's Wake, December 2006 <http://www.washingtonpost.com/wp-dyn/content/article/2006/12/26/AR2006122601257.html>

4. Stephen E. Lucas, 269.

5. Ibid, 270.

6. William Strunk Jr. and E.B. White, The Elements of Style, 4th ed., (Boston: Allyn and Bacon, 2000), 23-24.

7. Ibid, 18-19.

8. Ibid, 19-20.

9. Edward P. Bailey and Philip A. Powell, The Practical Writer, 7th Edition (Fort Worth, Texas: Harcourt Brace College Publishers, 1999), 50.

10. Merriam-Webster's Collegiate Dictionary, 10th ed., s.v. "argument."

11. Ibid, s.v. "reason."

12. Lynn Q. Troyka and Doug Hesse, Handbook for Writers, 8th ed. (Upper Saddle River, NJ: Pearson Education, Inc., 2007), 46.

13. Edward P. Bailey and Philip A. Powell, 98.

14. Stephen E. Lucas, The Art of Public Speaking, 6th Edition (Boston: McGraw Hill, 1998), 212.

15. Ibid, 212.

16. Merriam-Webster's Collegiate Dictionary, 10th ed., s.v. "objection."

17. Merriam-Webster's Collegiate Dictionary, 10th ed., s.v. "stage fright."

18. Stephen E. Lucas, 11.

19. Lilyan Wilder, 7 Steps to Fearless Speaking (New York: John Wiley & Sons, Inc, 1999), 246.

20. Ibid, 162.

21. John A. Kline, Speaking Effectively: A Guide for Air Force Speakers (Maxwell AFB: Air University Press, 1989), 47.

PHOTO CREDITS

154 White House photo

155 Ford: White House photo

155 Protesters: Julia Ryan, Haverford College (blog)

157 Library of Congress, via LOOK Magazine

158 U.S. Air Force photo

159 U.S. Air Force photo

166 Wikimedia Commons, under Creative Commons License

167 Building: St. Anselm College

174 White House photo

179 Time magazine, December 25, 2006

183 David Berry, used with permission

All photos of CAP cadets are from CAP sources, unless indicated otherwise.

INDEX

MCREL EDUCATIONAL STANDARDS

The *Learn to Lead* curriculum is correlated to Midcontinent Research for Education and Learning (McREL) standards for life skills, behavioral studies, career education, language arts, and civics. McREL maintains standards documents from professional subject area organizations and selected state governments. By referencing the McREL standards, the *Learn to Lead* curriculum demonstrates content relevance in the eyes of independent subject matter experts. For details, please see the *Learn to Lead Curriculum Guide* and capmembers.com/learntolead.

McREL standards are copyright 2010 by McREL

Mid-continent Research for Education and Learning
4601 DTC Blvd., Suite 500
Denver, CO 80237
Telephone: 303/337/0990
mcrel.org/standards-benchmarks

Used with permission.

LEARN TO **LEAD**
RESOURCES

A wealth of resources is available to support cadet leadership education. See capmembers.com/learntolead for free PDF editions, or purchase hard copies through Vanguard at CivilAirPatrolStore.com

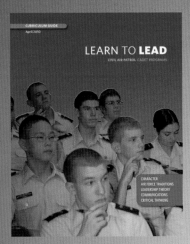

LEARN TO LEAD
CURRICULUM GUIDE

★ discusses the overall goals of cadet leadership education

★ outlines the content of all 4 volumes in the series

★ offers guidance on how to implement the curriculum

LEARN TO LEAD
ACTIVITY GUIDE

★ 24 hands-on team leadership problems

★ 6 movie discussion guides

★ 6 group discussion guides

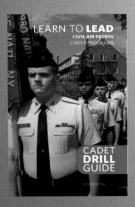

LEARN TO LEAD
CADET DRILL GUIDE

★ quick guidance on all major drill movements

★ tips on teaching using the demo-perf method

★ notes on basic formations

LEARN TO LEAD
LESSON PLANS

★ detailed, recipe-like lesson plans for volumes 1 & 2

★ available only online